Programming the iPhone User Experience

Programming the iPhone User Experience

Toby Boudreaux

O'REILLY®

Beijing · Cambridge · Farnham · Köln · Sebastopol · Taipei · Tokyo

Programming the iPhone User Experience
by Toby Boudreaux

Copyright © 2009 Toby Boudreaux. All rights reserved.
Printed in the United States of America.

Published by O'Reilly Media, Inc., 1005 Gravenstein Highway North, Sebastopol, CA 95472.

O'Reilly books may be purchased for educational, business, or sales promotional use. Online editions are also available for most titles (*http://my.safaribooksonline.com*). For more information, contact our corporate/institutional sales department: 800-998-9938 or *corporate@oreilly.com*.

Editor: Steven Weiss	**Indexer:** Seth Maislin
Production Editor: Sarah Schneider	**Cover Designer:** Karen Montgomery
Copyeditor: Emily Quill	**Interior Designer:** David Futato
Proofreader: Sarah Schneider	**Illustrator:** Robert Romano

Printing History:

August 2009:   First Edition.

ISBN: 978-0-596-15546-9

[M]

1249069284

Table of Contents

Preface ix

1. Cocoa Touch: The Core iPhone 1
   Mac Frameworks   1
      UIKit Overview   2
      Foundation Overview   4
   Garbage Collection   9
   The Devices   10

2. The Mobile HIG 11
   The Mobile HIG   12
   Enter Cocoa Touch   13
   Mobile HIG Concepts   13
      Provide One User Experience   13
      Provide Seamless Interaction   15
      Let the User Know What's Going On   16
      Use Progressive Enhancement   16
      Consider Cooperative Single-Tasking   17
   A Supplement to the HIG   18

3. Types of Cocoa Touch Applications 19
   Productivity Tools   20
      Limited or Assisted Scrolling   20
      Clear and Clean Detail Views   23
   Light Utilities   24
   Immersive Applications   25

4. Choosing an Application Template 27
   View Controllers   29
      View Controller Subclasses and Corresponding Application Templates 30
   Core Data Templates   35

5. Cooperative Single-Tasking **37**

Task Management and iPhone OS 37
    Example Application 38
Launching Quickly 43
    Example Application 45
Handling Interruptions 47
    Interruptions and the Status Bar 48
    Example Application 48
Handling Terminations 51
    Example Application 51
Using Custom URLs 52
Using Shared Data 54
Using Push Notifications 55

6. Touch Patterns **57**

Touches and the Responder Chain 58
    UITouch Overview 58
    The Responder Chain 59
Touch Accuracy 62
    Size 62
    Shape 66
    Placement 67
    Overlapping Views 68
Detecting Taps 68
    Detecting Single Taps 68
    Detecting Multiple Taps 69
Detecting Multiple Touches 70
Handling Touch and Hold 70
Handling Swipes and Drags 72
Handling Arbitrary Shapes 74

7. Interaction Patterns and Controls **83**

Application Interaction Patterns 83
    Command Interfaces 83
    Radio Interfaces 84
    Navigation Interfaces 85
    Modal Interfaces 85
    Combination Interfaces 87
UIControl Classes 88
    The Target-Action Mechanism 89
    Types of Control Events 89
Standard Control Types 91
    Buttons 91

Modal Buttons   98
Sliders   103
Tables and Pickers   106
Search Bars   109
Segmented Controls   111
Scrolling Controls   114
Tables and Embedded Controls   120
Passive Indicators   121
Active Indicators and Control Accessories   122

8. Progressive Enhancement **125**
Network Connectivity   126
Maintain State and Persist Data   126
Cache User Input   127
Reflect Connectivity Appropriately   128
Load Data Lazily   129
Peer Connectivity with GameKit   132
Location Awareness   133
Accelerometer Support   137
Rotation Support   139
Audio Support   140

9. UX Anti-Patterns **147**
Billboards   147
Sleight of Hand   150
Bullhorns   152
App As OS   155
Spin Zone   157
The Bouncer   157
Gesture Hijacking   160
Memory Lapse   161
The High Bar   163
Sound Off   164

Index **167**

Preface

The launch of the iPhone software development kit (SDK) was a big deal for developers, designers, and consumers alike. Developers and designers were able to access a previously closed platform and distribution channel. Consumers were excited to explore an endless stream of new applications created by passionate independent developers and innovative companies.

New platforms often suffer from growing pains. Users and application creators learn simultaneously, with developers releasing applications to the market and users providing feedback. Different application teams come up with different approaches to common problems, because agreed-upon, proven solutions take time to emerge. These growing pains can be compounded when communication within the community is minimal.

In the case of the iPhone SDK, Apple has famously frustrated both developers and consumers by imposing a non-disclosure agreement (NDA) that legally restricts the ability to discuss upcoming features, tools, approaches, and technologies. To compensate for the lack of conversation within the development community, Apple provides a great set of guidelines for designing and coding iPhone applications.

The Human Interface Guidelines (HIG) describe the way applications should look and feel on Apple platforms. For the iPhone OS, Apple released a separate version of the HIG that focuses on mobile, Multi-Touch applications. The HIG works well in many regards, and it remains a valuable resource for anyone creating mobile applications.

However, the HIG cannot cover all topics that arise in the course of application development, nor can it provide insight from the market at large. Questions invariably emerge: What works for users? What causes frustration? What habits have emerged that should be avoided? What practices can help small teams or independent developers use their limited time and resources wisely? Which features should be prioritized for a shipping product? What do programmers need to know to deliver a great user experience?

You can think of this book as a supplement to the HIG—a resource that, along with Apple's extensive technical documentation, should guide teams through the choices they must make when embracing a new platform.

Audience for This Book

This book is geared toward designers, developers, and managers who wish to develop user-friendly applications for the iPhone and iPod Touch. The book mixes technical and strategic discussions, and it should be approachable by both technical developers and technology-savvy users.

The code in this book is Objective-C, and an understanding of the language is necessary to maximize the value of the code examples. If you are a desktop Cocoa developer, this book will introduce you to the differences between Cocoa and Cocoa Touch, the set of frameworks for iPhone applications. Managers and experience designers can use this book to understand the ways that applications can function together to create a holistic user experience.

Finally, this book is for readers who own and use the iPhone. To create an excellent iPhone application, a developer must have empathy for iPhone users. An appreciation of the challenges that face mobile users—both environmental and physical—is essential.

If you have no prior experience with Objective-C or Cocoa Touch, you may want to refer to an excellent book by one of this book's technical editors, Jonathan Zdziarski. His book, *iPhone SDK Application Development* (O'Reilly), provides a technical foundation in Objective-C and Cocoa Touch.

Organization of This Book

Chapter 1, *Cocoa Touch: The Core iPhone*, describes the essential information for Cocoa Touch and the devices that run the iPhone OS.

Chapter 2, *The Mobile HIG*, gives an introduction to the Human Interface Guidelines and elaborates on the most important concepts in the iPhone user experience.

Chapter 3, *Types of Cocoa Touch Applications*, presents a vocabulary for describing families of applications for the iPhone and links each to a structural application type.

Chapter 4, *Choosing an Application Template*, examines the application templates supplied with Xcode and the iPhone SDK. The concept of view controllers is explained with each type of standard view controller covered.

Chapter 5, *Cooperative Single-Tasking*, breaks from the application structure and focuses on the ways applications can work together to create a holistic user experience.

Chapter 6, *Touch Patterns*, teaches you how to work with the Multi-Touch interface, including design patterns for standard and custom gestures.

Chapter 7, *Interaction Patterns and Controls*, covers the types of user interface controls included in the Cocoa Touch UI framework, and the design patterns used to enable controls to work together.

Chapter 8, *Progressive Enhancement*, discusses techniques to layer functionality around user ability. Networking, data management, rotation, and audio functionality are addressed.

Chapter 9, *UX Anti-Patterns*, covers a set of common approaches that can cause issues for users.

Conventions Used in This Book

The following typographical conventions are used in this book:

Italic
> Indicates new terms, URLs, email addresses, filenames, and file extensions.

Constant width
> Used for program listings, as well as within paragraphs to refer to program elements such as variable or function names, databases, data types, environment variables, statements, keywords, classes, and frameworks.

Constant width bold
> Used for emphasis within code examples.

 This icon signifies a tip, suggestion, or general note.

Using Code Examples

This book is here to help you get your job done. In general, you may use the code in this book in your programs and documentation. You do not need to contact us for permission unless you're reproducing a significant portion of the code. For example, writing a program that uses several chunks of code from this book does not require permission. Selling or distributing a CD-ROM of examples from O'Reilly books does require permission. Answering a question by citing this book and quoting example code does not require permission. Incorporating a significant amount of example code from this book into your product's documentation does require permission.

We appreciate, but do not require, attribution. An attribution usually includes the title, author, publisher, and ISBN. For example: "*Programming the iPhone User Experience* by Toby Boudreaux. Copyright 2009 Toby Boudreaux, 978-0-596-15546-9.*"

If you feel your use of code examples falls outside fair use or the permission given above, feel free to contact us at *permissions@oreilly.com*.

Safari® Books Online

When you see a Safari® Books Online icon on the cover of your favorite technology book, that means the book is available online through the O'Reilly Network Safari Bookshelf.

Safari offers a solution that's better than e-books. It's a virtual library that lets you easily search thousands of top tech books, cut and paste code samples, download chapters, and find quick answers when you need the most accurate, current information. Try it for free at *http://my.safaribooksonline.com*.

How to Contact Us

Please address comments and questions concerning this book to the publisher:

> O'Reilly Media, Inc.
> 1005 Gravenstein Highway North
> Sebastopol, CA 95472
> 800-998-9938 (in the United States or Canada)
> 707-829-0515 (international or local)
> 707-829-0104 (fax)

We have a web page for this book, where we list errata, examples, and any additional information. You can access this page at:

> *http://www.oreilly.com/catalog/9780596155469*

To comment or ask technical questions about this book, send email to:

> *bookquestions@oreilly.com*

For more information about our books, conferences, Resource Centers, and the O'Reilly Network, see our website at:

> *http://www.oreilly.com*

Acknowledgments

I would like to thank Steve Weiss and Robyn Thomas, my editors at O'Reilly Media, for their guidance throughout this project. I would also like to thank Chandler McWilliams and Jonathan Zdziarski for their careful technical reviews, which made this a more accurate and comprehensive book. Finally, thanks to my wife and son for their flexibility and support, and for sacrificing weekend trips as I worked on the project.

CHAPTER 1

Cocoa Touch: The Core iPhone

Cocoa is a collection of tools—libraries, frameworks, and APIs—used to build applications for the Mac OS. Most of the core functionality you would need to develop a rich Mac application is included in Cocoa. There are mechanisms for drawing to display, working with text, saving and opening data files, talking to the operating system, and even talking to other computers across a network. The look and feel of Mac applications is recognizable and relatively consistent in large part because of the breadth and quality of the Cocoa user interface framework.

The Cocoa frameworks include two areas of focus: classes that represent user interface objects and collect user input, and classes that simplify challenges like memory management, networking, filesystem operations, and time management.

Developing applications for the iPhone and iPod Touch is similar in many ways to building applications for Mac OS X. The same tools are used for writing and debugging code, laying out visual interfaces, and profiling performance, but mobile application development requires a supplemental set of software libraries and tools, called the iPhone SDK (software development kit).

Cocoa Touch is a modified version of Cocoa with device-specific libraries for the iPhone and iPod Touch. Cocoa Touch works in conjunction with other layers in the iPhone and iPod Touch operating systems and is the primary focus of this book.

Mac Frameworks

Mac OS X programmers use a framework called `AppKit` that supplies all the windows, buttons, menus, graphics contexts, and event handling mechanisms that have come to define the OS X experience. The Cocoa Touch equivalent is called `UIKit`. In addition to user interface elements, `UIKit` provides event handling mechanisms and handles drawing to the screen. `UIKit` is a very rich framework and is a major focus of user experience programmers. Nearly all user interface needs are accounted for in `UIKit`, and developers can create custom UI elements very easily. Many of the user experience problems and patterns addressed in this book will focus on `UIKit` programming with an emphasis on standard solutions.

The second Cocoa Touch framework is the `Foundation` framework. You can think of `Foundation` as the layer that abstracts many of the underlying operating system elements such as primitive types, bundle management, file operations, and networking from the user interface objects in `UIKit`. In other words, `Foundation` is the gateway to everything not explicitly part of the user interface. As you'll see in this book, user experience programming goes deeper than the user interface controls and includes things such as latency management, error handling, data caching, and data persistence.

UIKit Overview

The user interface comprises the elements of a device or application that users see, click, and—in the case of Cocoa Touch—tilt, shake, or tap. User interfaces are a big part of user experience. They provide the *face* of your product, and often a bit more.

For the most part, `UIKit` is just a limited subset of the `AppKit` framework for Mac OS X. If you have experience developing Cocoa apps for the Mac, you will get your head around `UIKit` fairly quickly. The main differences are that `UIKit` is tuned for specific hardware interfaces and that it provides less functionality than `AppKit`. The reduced scope of `UIKit` is primarily due to the differences in robustness between typical computers and the iPhone or iPod Touch. Despite the omission of a few familiar elements, `UIKit` is a very capable toolset.

The best way to understand the breadth of `UIKit` is with a visual topology of the framework. Figures 1-1 and 1-2 show the layout of `UIKit`.

The core class from which all Cocoa objects inherit basic behavior is `NSObject`. The `NS` prefix has roots in the non-Apple origins of Cocoa at NeXT. The early versions of what is now Cocoa were called NextStep. Most Cocoa classes in Cocoa are subclasses of `NSObject`, and many classes assume that `NSObject` is the foundation of objects being passed around. For example, the class `NSArray`, which represents a collection of pointers, requires that any pointer it stores points to an `NSObject` subclass. In most cases, any custom class you create should inherit from `NSObject`.

The names of classes in Cocoa are often very descriptive. The following illustrations give an overview of the classes in `UIKit`. The purpose is to provide a view of the entire collection of classes so that developers of all experience levels can see the breadth of the frameworks.

In `UIKit`, all classes that respond to user input inherit from `UIResponder`, which is an `NSObject` subclass that provides functionality around handling user input. Figure 1-1 focuses on the subclasses of `UIResponder`.

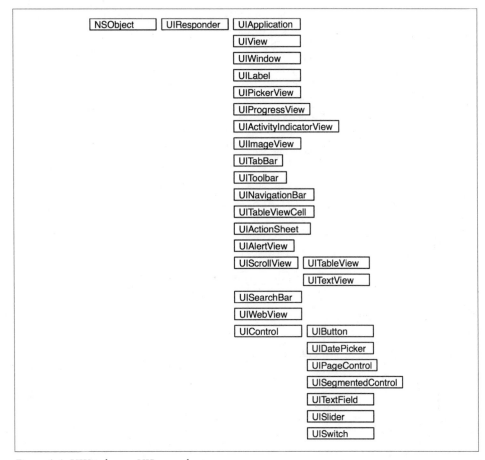

Figure 1-1. UIKit classes: UIResponder tree

In addition to the UIResponder class hierarchy, UIKit includes a set of classes acting as value objects, logical controllers, and abstractions of hardware features. Figure 1-2 shows these classes.

The documentation sets that accompany the iPhone SDK, in tandem with the Apple Developer Connection website (*http://developer.apple.com/*), cover all the details of the Cocoa and Cocoa Touch framework classes. This book will further elaborate on key classes in the context of interaction design patterns and overall user experience, but the official documentation should remain the primary reference for developers, as each update to the iPhone SDK includes amended or edited documentation packages.

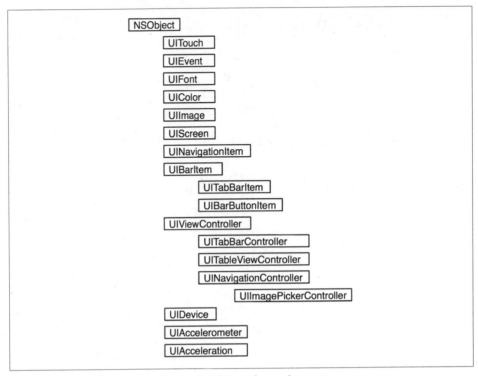

Figure 1-2. UIKit classes: controllers, value objects, device classes

Foundation Overview

The Foundation layer of Cocoa Touch (and Cocoa on the Mac) provides an object-oriented abstraction to the core elements of the operating system. Foundation handles core features of Cocoa Touch, including:

- Essential object behavior, such as memory management mechanisms
- Inter-object notification mechanisms, such as event dispatching
- Access to resource bundles (files bundled with your application)
- Internationalization and localization of resources, such as text strings and images
- Data management tools (SQLite, filesystem access)
- Object wrappers of primitive types, such as NSInteger, NSFloat, and NSString

All Cocoa Touch applications must link against Foundation because Foundation contains the classes that make a Cocoa application work—including many classes that are integral in the functioning of the user interface framework. For example, many UIKit methods use NSString objects as arguments or return values from methods.

The Foundation class tree as supported by Cocoa Touch is illustrated in Figures 1-3 to 1-9. The class hierarchy diagrams are logically grouped according to coarse functionality. This conceptual grouping mirrors the organization used by Apple in the *Cocoa Fundamentals Guide* included in the developer documentation. You should consult the *Cocoa Fundamentals Guide* and the class documentation provided by Apple as part of the iPhone SDK install for updated, in-depth information about the framework classes.

Figure 1-3 shows a subset of NSObject subclasses that represent value objects. Value objects are used to represent non-functional values—primitives, dates, generic binary data—and to provide utility methods for those values.

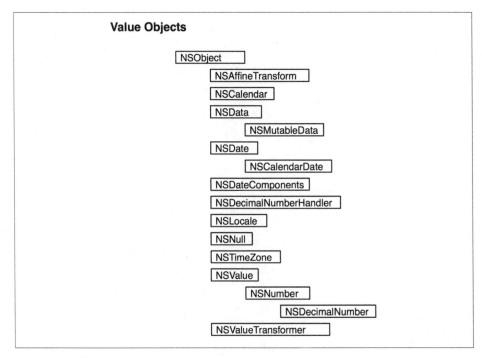

Figure 1-3. Value objects

Figure 1-4 maps the NSObject subclasses that focus on XML and string management. The XML classes are particularly useful when working with web services. Strings are used constantly, and Cocoa developers spend a lot of time working with NSString instances.

Foundation provides powerful classes for collection management. These classes are shown in Figure 1-5. Standard collection types such as arrays, sets, and hash tables are included, along with classes used for enumerating through collections.

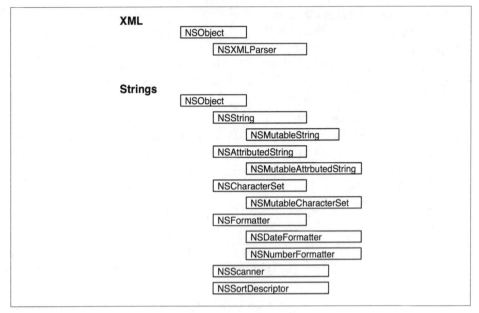

Figure 1-4. XML and strings

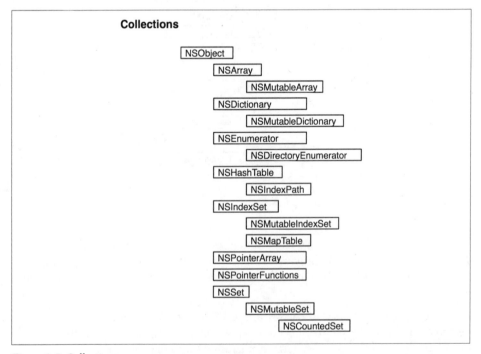

Figure 1-5. Collections

Figure 1-6 illustrates classes that focus on operating system services, file operations, and inter-process communication (IPC).

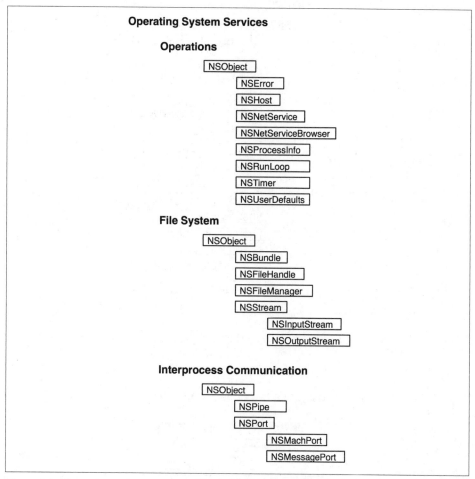

Figure 1-6. Operating system services: operations, file operations, interprocess communication

Figure 1-7 illustrates the portion of Foundation that provides for URL handling functionality. Hybrid web/Cocoa applications use URL handling classes heavily.

Figure 1-8 shows the classes used to manage threading in Cocoa applications. Careful thread management can be an important part of optimizing the perception of responsiveness in an application.

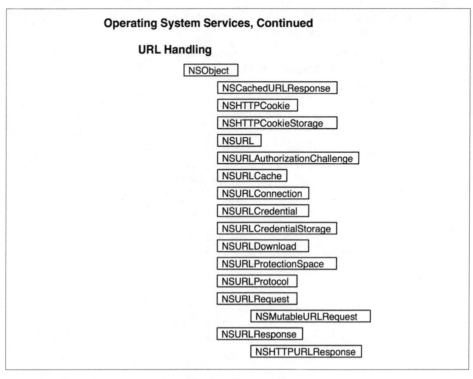

Figure 1-7. Operating system services: URL handling

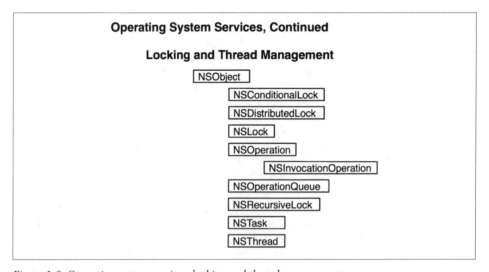

Figure 1-8. Operating system services: locking and thread management

Finally, Figure 1-9 shows classes providing notifications, archiving, and core language features such as exceptions.

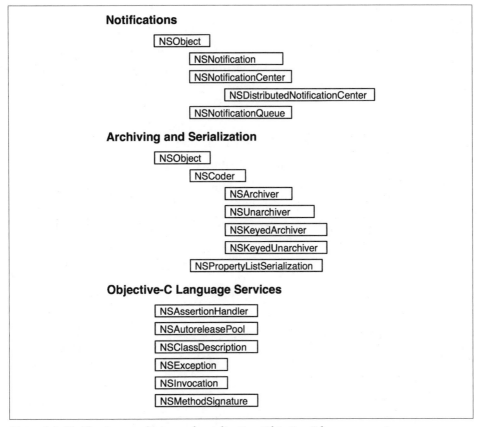

Figure 1-9. Notifications, archiving and serialization, Objective-C language services

Garbage Collection

One notable difference between the iPhone OS and Mac OS X is that the implementation of Foundation for Cocoa Touch does not automatically recover memory when objects are destroyed. This means developers must keep track of the objects they create and follow certain idioms in order to keep memory usage as low as possible.

The newest version of the Objective-C programming language, version 2.0, added support for automatic resource management, or *garbage collection*. Developers who have grown accustomed to using garbage collection in Cocoa applications on the Mac may find its omission in the iPhone SDK inconvenient. The performance implications of most garbage collection implementations are important considerations on mobile devices, and this is a key reason it was excluded from the iPhone. Battery life and processing speed are important elements of an elegant user experience, and often act as differentiating factors among competing devices in the consumer marketplace.

The Devices

The screen on both the iPhone and iPod Touch is an LCD-lit 3.5-inch (diagonal) widescreen Multi-Touch display. The screen resolution is 480 × 320 pixels at a resolution of 163 pixels per inch. The devices include the following sensors: accelerometer, proximity sensor, and ambient light sensor.

- When activated, the *accelerometer* is used to detect device movement in space, providing information about movement along three axes.
- The *proximity sensor* recognizes the proximity of the handset to another object, most commonly a human ear.
- The *ambient light sensor* detects the level of ambient light hitting the device.

Both the iPhone and iPod Touch devices provide rocker switches for controlling volume, a hardware power button, and a depressible "home" button. These concrete interfaces are outside the scope of Cocoa Touch programming, but are notable in the overall UX (user experience) of the devices.

From the point of view of user experience programmers, the hardware interface elements are separate from the touch interface. Apple doesn't provide any means of accessing the home button, lock button, volume controls, or the navigation controls included on headsets. This simplifies the domain of UX programming, but comes at a cost: there are certainly cases in which access to the home button or volume rocker could provide enhanced functionality to users.

The iPhone provides a few distinct advantageous features over the iPod Touch, aside from the telephony. For example, the iPhone includes GPS support and a hardware ringer/silence switch. For networking, in addition to 802.11g support, the iPhone 3G includes support for 3G wireless networking, and support for High-Speed Downlink Packet Access (HSDPA). The operating system will, unless told otherwise, automatically switch from 3G to wireless networks with known names. This feature is part of a larger UX pattern that is core to the iPhone: attempting to deliver the best experience possible given the environment, without interruption. We will cover this pattern in Chapter 8.

CHAPTER 2

The Mobile HIG

Most large software efforts—especially those allowing any form of extension by developers—define guidelines for user experience. These guidelines provide documentation of the design, interaction, and semantic patterns that define the interaction between humans and the software in question.

Apple is known for compelling, forward-thinking user experiences. Their tools and libraries make the creation of third-party software that fits seamlessly into the aesthetics of the Mac OS X operating system a trivial task. The Mac "look and feel" is something users recognize and expect from the applications. Apple provides developers and designers with a set of general Human Interface Guidelines (HIG) to help clarify their approach and reasoning behind interface decisions.

There has almost always been controversy around the Apple HIG, leading some independent developers to proclaim the Apple HIG a "dead" document. Most of this has been due to Apple stepping outside their own recommendations and guidelines, and has thus created three tiers of applications: those by Apple, those by developers who follow the HIG, and those by developers who ignore the HIG (think Java and Swing applications).

The benefits of designing within the boundaries of the HIG are significant for both customers and developers. Users can learn to interact with an application much faster when the design of the interface follows familiar conventions. The Mac look and feel is skewed toward first-time or casual users. For frequent power users, *progressive enhancement* techniques are used to add options and customization without alienating newcomers. When done properly, progressive enhancement adds depth, rather than breadth, to the user experience.

Software producers benefit in many ways as well. Development is quicker because a rich set of standards can allow developers to focus on the unique elements of an application instead of fretting over and excessively prototyping common layouts and visual effects. The same is true for decisions about features. The HIG describes a layered structure for prioritizing functionality and design. When making design decisions or focusing on implementation, the recommendation is to focus first on the minimum requirements for your application. Next, add features that users expect from the application, including things like keyboard shortcuts, preference management, and undo support, along with modern Cocoa interfaces. The final, lowest-priority layer should be differentiation from similar applications. Attempts at differentiation add risk to projects but often result in progressive, beautiful software and happy customers.

With every release of Mac OS X, Apple provides additions to the toolkit and often updates to the HIG. Occasionally, the updates are retroactive, incorporating articulations of patterns and UI enhancements already added to production software. Even for skeptics, the HIG has remained an important touchstone when thinking of user experience on Mac OS X.

The Apple HIG includes:

- Specifications for UI elements, such as buttons
- Use cases for all user input, such as mouse clicks
- Consistency across disparate applications
- Exception and error handling conventions
- Conventions for prompting users for input
- Conventions for displaying interrupts to users
- Latency feedback patterns and indicators
- Compound control events, such as using modifier keys

The Mobile HIG

There are many of us in the Apple developer community who hope that Cocoa Touch will extend beyond the two current mobile devices on which it is implemented. Currently, all Apple laptops support Multi-Touch input in a limited fashion, allowing application-specific gestures for zooming and rotating views. Still, Cocoa Touch is being positioned as a mobile platform, as is evident from the title of the new HIG: *iPhone Human Interface Guidelines*, sometimes referred to as the mobile HIG.

Naturally, Apple is simply keeping focus where it should be: developing applications for known, released devices and operating systems. In the future, I hope that much of what is covered in this book and in the mobile HIG will be applicable to development for laptops, desktops, tablets, and any new devices Apple releases.

In a sense, this book functions as a supplement to Apple's mobile HIG. I will expand on many of the points in the HIG, giving example implementations of patterns and concepts, and citing examples using apps you probably own and use.

Enter Cocoa Touch

The introduction of Cocoa Touch was important to developers and experience designers not only because it meant new hardware interfaces, but also because it signified an expansion of Apple's thought into new interaction patterns. Touch, Multi-Touch, orientation, acceleration, and gestural interaction patterns were not new to the world when Apple announced Cocoa Touch. Still, nobody had approached touch-based interaction with a comprehensive, user-focused vision until the development of Cocoa Touch.

Naturally, as with all interaction innovations, Apple needed to provide an update to the HIG for its touch framework. The company realized that the nature of Cocoa Touch applications was different from that of standard Cocoa apps, and though it worked very hard to maintain consistency between the two, Apple decided to release a separate human interface guidelines document for Cocoa Touch.

Mobile HIG Concepts

The mobile human interface guidelines are described in a large, detailed, useful document called *iPhone Human Interface Guidelines* and are included in the iPhone SDK documentation. As with interface guidelines for any platform, you should know the HIG inside and out so that you can take the path of least resistance where such a path exists. Try to avoid breaking the guidelines for market differentiation or other reasons that aren't user-centered. Instead, have faith in the expectations of the audience, and use pricing, marketing efforts, and a focus on advanced and valuable details to one-up your competitors.

One warning: Apple controls the single distribution channel for applications and reserves the right to reject any application from the App Store for any reason. Unless you're developing applications for hacked devices, the App Store is the only means of distributing an application to a market. When submitting an application, you must agree that your application adheres to the mobile HIG. There are countless examples of applications that eschew the HIG in some respect but still make it into the store. Conversely, there are at least a few well-known cases in which rejections have been based solely on nonconformance with the HIG. Break the rules at your own peril, and choose your battles wisely without giving up on a compelling user experience.

Provide One User Experience

The launch of the iPhone SDK was a keystone moment for many types of developers. There were large communities of developers with expert knowledge of Cocoa, web, and mobile programming, but nobody had experience with the iPhone as a platform. Given that the iPhone SDK includes elements that cross all these disciplines, and that the platform launched as a brave new world, there was a high potential for a "wild west" sort of user experience. For a company focused on UX as a key differentiator, and users accustomed to consistent, beautiful devices and applications, the release of a heavily hyped SDK for a massively popular new device would likely yield applications that competed for attention, leaving users with feature fatigue.

If you take a fresh look at the iPhone with an eye on UX, a few important attributes stand out:

- The hardware is designed to be unobtrusive to the software. The display is as large as technically practical, with high fidelity and no seams or edges. There are very few buttons or switches, allowing users to focus on the display.

- The lighting (when enabled) adjusts to the user's environment, allowing the device to blend into the background and keep the screen contents consistently visible and in focus.

- There is no branding to distract from or compete with the current application.

- The shape is sculpted to allow easy retrieval and storage in a pocket—the expectation is for users to visit, remove, and revisit focus on the device as needed.

- The Home screen is immediately visible, with no interstitial distractions such as splash screens.

- The full state change between application screens or pages instead of partial scrolling establishes that interaction should be visceral but imprecise. Intent is more important than accuracy.

- The dock tray provides four slots for users to fill with their most frequently accessed applications. This simple design gives users the ability to prioritize applications with the greatest utility, keeping them only one touch away at all times.

Many of these attributes focus on and strengthen the most important UX rule in the world of Cocoa Touch: applications should be a part of a single user experience modeled for a person with a powerful mobile device and many disparate, specialized, but related needs. Applications should not typically create terribly distinct user experiences. That simplification sounds a bit extreme, so it's important to note that applications should have their own identity. It would be insanity to believe that all problems are the same for all people, and that all design patterns apply to all problems equally.

In paying attention to the details of UX programming, there are many points at which interaction decisions can be made. Every application should find balance between invisibly fitting into the whole experience and providing its own value and uniqueness.

In this book, and in the mobile HIG, a guiding mantra is this: develop an application that can be accessed as often as needed, as quickly as possible, and that will solve the task at hand in cooperation with the entire system.

Provide Seamless Interaction

Mobile devices are used under highly variable conditions. A mobile device can have amazing value to a user on the go: at the gym, on a commuter train, or while traveling. The value of mobile Internet access beyond the confines of home and office is significant—even more so when the barrier to access is very low and the adaptability to environment is very high.

Under the ethic of cooperatively providing utility to users in a streamlined fashion, developers and designers can explore certain key points in the HIG, and add their own:

- Splash screens are evil. While branding is important, the proper place for it is in the iconography, optional "About" or "Info" screens, and App Store profiles. The most common interaction pattern with iPhone applications is to launch them frequently, close them quickly, and treat them as part of a set of tools that interact to comprise a single user experience. Splash screens break the perception of seamlessness.

  The HIG offers a very useful suggestion for managing launch states, which may be quite slow, depending on the needs of your application. The suggestion is to provide a PNG image file in your application bundle that acts as a visual stand-in for the initial screen of your application. For example, if the main screen for your application is a table full of data, provide an image of a table without data to act as a stand-in. When your data is ready to be displayed, the image will be flushed from the screen, and the user experience will feel more responsive.

  In this book, we will explore extensions of this, including a pattern for loading application state lazily.

- Speed is king. Your application should launch quickly and smoothly. It should also close quickly and smoothly. Users should feel more like they are pausing and unpausing an application rather than starting and quitting.

- Consider state maintenance. There are many reasons an application might terminate, and not all of them are user-controlled. Your application will receive a message from the operating system letting it know that it will be terminated. This gives you an opportunity to improve the feel of your application's responsiveness by selectively taking a snapshot of the state of your application and persisting it for the next launch. Have your application detect whether the user has just filled a text box with the next great American novel before simply releasing that data and exiting.

- The standard icon size for Cocoa Touch applications is 57 pixels × 57 pixels. On the screens of the current round of devices, this is approximately four-fifths of an inch (0.8 inches). The spacing of the applications on the Home screen, combined with the icon dimensions, sets the stage for touch fidelity. Though there are certainly very usable applications that require high fidelity, you should always keep in mind the size of fingertips, the conditions for use of applications (in cars, on trains, while walking), and the frustration that might result from an expectation of precision that is beyond reason.

It's not difficult to imagine scenarios where a need for precision works against common interaction patterns. As with many UX considerations, this problem is just one of many to keep in mind when designing, gathering user feedback, and providing logical enhancements to input mechanisms.

- Modifying actions, such as creating, updating, or deleting an object, are common on a platform focused on utility. Users should never be left in the dark in regards to a network, filesystem, or object manipulation. The touch metaphor is quite literally about the visceral nature of object manipulation (translated to a 2D surface). Keep in mind Newtonian rules when performing modifications. That is, every action has an equal and opposite reaction. The next section covers this concept in more detail.

Let the User Know What's Going On

A heavy portion of the marketing materials for the iPhone and iPod Touch focuses on "having the Internet in your pocket," and it's not unreasonable to assume that the networking abilities of the devices are key selling points. The Internet at large has worked its way into the fabric of our daily lives, and Internet-enabled devices provide incredible portals into that connectivity. It's vital to consider all outcomes when developing networked applications—both for users and network nodes, such as mobile devices. When handling network communications, take the time to explore and provide handling for all success and failure states described in the protocols you'll be supporting.

Filesystem IO, such as saving to databases and local files, is an example of local, but unpredictable, interactions. For example, database inserts can fail and files might be missing. Though many operations are possible within a single pass through the event loop, it's important to give an indication that something is about to happen, is happening, or has completed, as appropriate to the expected duration of operations at hand. More casually, if someone is saving a record, let them know the outcome. If they're searching for records and the operation moves beyond a single pass through the event loop, let them know how the search is proceeding. While there is a balance between starving users for information and overwhelming them, most users would prefer to know the status of an operation they've initiated.

Use Progressive Enhancement

On the subject of networking and the unpredictable conditions under which Cocoa Touch applications are used, an exploration of progressive enhancement is worthwhile. Progressive enhancement is a strategy for layering features so that a minimum baseline is always present, and additional secondary features or design elements are added when conditions allow.

Progressive enhancement is an important concept in developing user experiences for any portable platform. This includes technical variation, such as network availability and speed, but also user variation, such as temporary or permanent abilities of users. For example, color blindness might be an important concern for information designers. Expected ability to focus might matter considerably for a navigation application.

It's actually hard to imagine a non-immersive application without use cases that include variations in audience skill, physical ability, or interest level. Interaction with networks and computers is another common requirement for iPhone applications. Covering a baseline case for each requirement and adding value in capable contexts is a great approach to making users happy.

The book will explore progressive enhancement in detail in Chapter 8, *Progressive Enhancement*, but for now, consider the following problems:

- Networked data is subject to throughput, consistency, and reachability. Applications should attempt to provide some value in the presence of even the spottiest networks.
- You may prefer that users interact with your application in very controlled environments, allowing intense focus and deliberation. Still, it is worth planning for use on jarring subway trains or in cars.
- Adhering to the mobile HIG should theoretically yield applications with very shallow learning curves, but it's inevitable that any application providing new value for users will include its own distinct behaviors. Where possible, attempt to provide nearly as much value for users the first time they launch your application as for those launching it 20 times over the course of a day. Small barriers to entry add up.

Consider Cooperative Single-Tasking

If each application fills a relatively small but important need in a complex mobile lifestyle, a question arises about the holistic user experience. How can applications work together to create a complete toolset for a mobile user?

The aspect of applications working together to provide a single, holistic experience can be called *cooperative single-tasking*.

Cocoa Touch applications can take advantage of a design pattern you already know from the Internet: URIs with custom protocol handlers. Two simple examples of this behavior use the `mailto://` and `http://` links to launch Mail and Safari, respectively.

It is a fairly simple process to register your own schemes inside the application property list file and to provide methods in your application delegate to process these URIs.

A Supplement to the HIG

As you read and refer to this book, consider it a companion document to the official Apple HIG. You will find most of the concepts in the HIG discussed here, along with additional examples, best practices, and challenges. This book doesn't follow the structure of the HIG, but the concepts should be easy to reference. Together, the two should provide developers and designers with the technical and conceptual information required to create applications that feel at home on the iPhone and iPod Touch.

CHAPTER 3

Types of Cocoa Touch Applications

Cocoa Touch applications tend to fit one of several key types. Like most classification systems, the list in the mobile HIG that is elaborated upon here is far from perfect. We can all think of applications that fit into multiple categories, or new categories presently missing from the listing. Think of this less as a strict classification system and more as a list of general directions in which an application developer might lean. Treat the terms as indicators for developer intent and market positioning.

The application types are as follows:

- Productivity tools
- Lightweight utilities
- Immersive media

You can consider this chapter as a supplement to the information Apple provides in the mobile HIG. Before elaborating on each application type, it's worth providing a secondary classification schema: the one that the Apple App Store uses.

The App Store files applications into one of the following categories. This list of categories is editorial in nature and is likely to change as new applications are released into the store. Still, keep this organization in mind when developing your user experience. The App Store is the only official distribution channel for Cocoa Touch applications and thus, in a sense, your user experience officially begins in the App Store.

Books	News
Business	Photography
Education	Productivity
Entertainment	Reference
Finance	Social Networking
Games	Sports
Healthcare & Fitness	Travel
Lifestyle	Utilities
Music	Weather
Navigation	

Productivity Tools

Productivity-focused applications are those that provide organizational benefits and streamlined access to detailed data. Mobile productivity apps tend to be used heavily, as mobile users are by definition "on the go" and thus consult their devices in short bursts to accomplish very specific tasks. When making UX decisions for productivity tools, it's vital to consider the preferences of the task-oriented user persona.

According to the mobile HIG, the most common UX tasks focus on managing and engaging hierarchical graphs of information, often represented as tree-like data structures (similar to a family tree). An example of a productivity-focused iPhone application is the built-in Contacts application. Contacts provides a very accessible list of people and personal information (email addresses, phone numbers, physical addresses, birth dates) stored on your iPhone. The application interface is based on a well-established pattern in which a table UITableView displays information at the highest point in the graph. When a user selects a row in the table, a detailed view of the row shows on a new screen that slides in from the right.

Figure 3-1 shows the left-to-right interface pattern for moving down a tree-like graph toward more specific and detailed information.

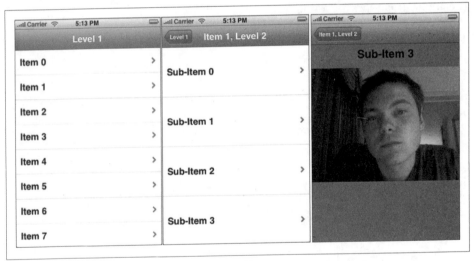

Figure 3-1. Drilling down through a graph using table views

Limited or Assisted Scrolling

In general, try to reduce the need for vertical scrolling in your application unless the nature of the content in the view (such as an email message or news article) is too large to display succinctly. Despite the elegant animations and intuitive swipe patterns supported by native view (UIView) objects, Cocoa Touch applications should strive for limited content outside the main viewing area. Naturally, scrolling is necessary in many cases, especially when dealing with a wide set of nodes representing large groups of siblings. For example, consider the graph in Figure 3-2, which represents a library of books, their subjects, and their authors.

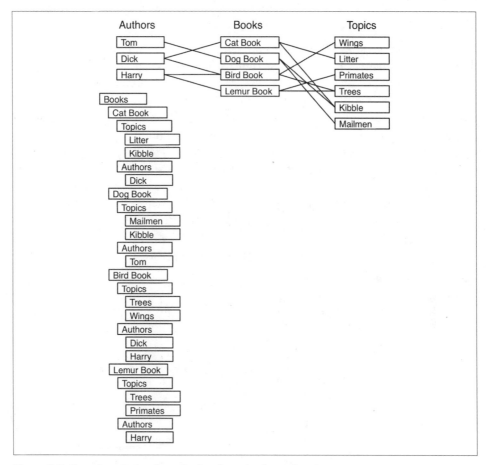

Figure 3-2. Complex relational graph of authors, books, and topics

Because of the many-to-many relationships between authors and books, books and subjects, and, as a transitive function, authors and subjects, there are many ways to present the data. The presentation depends on the expectations and needs of the audience. By focusing on the most pressing needs of a given audience, you can eliminate extraneous information and reduce complexity—two excellent goals for Cocoa Touch designers.

When scrolling is a requirement, it's far from the end of the world. Limiting the need for interaction can be helpful, but the iPhone is a platform meant to facilitate getting things done, and thus supports scrolling very well. When handling very large, structured lists, you can take advantage of one-touch index browsers. This assisted scrolling helps to provide the kind of functionality we're all used to from standard desktop computers—namely, the ability to "page down" with one touch rather than having to manually swipe repeatedly to scroll the views.

The `UITableView` class in `UIKit` provides a nice assistive feature for browsing larger datasets. If you organize your table according to a logical scheme such as alphabetical sorting, you can add a touchable vertical ribbon to the right of the view that lets users jump to particular points in the index.

Figure 3-3 shows an index ribbon for an alphabetical list. Touching any letter in the index ribbon causes the table to scroll to the section corresponding to the selected index value. Note that alphabetization is just a simple example—the index section shortcuts can be any string you choose. Figure 3-4 shows a table view indexed by month name.

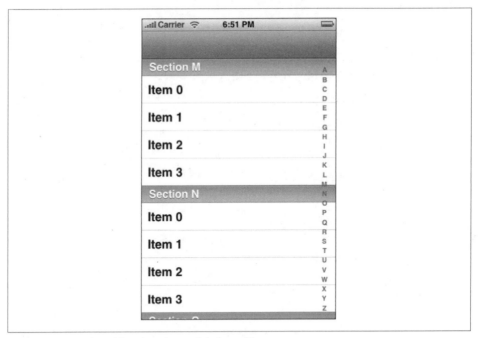

Figure 3-3. An index ribbon for a large alphabetical list

Figure 3-4. An alternate index ribbon format

Clear and Clean Detail Views

When designing a view that represents a content object (such as a person in the Contacts application), it can be tempting to display more information than a user needs. Obviously, overloading users with information isn't nearly as helpful as using focused design to guide them along a specific path.

Detail views designed so that important information can be presented on a single screen without scrolling or panning increases usability because less interaction is required. A key challenge is to balance this desire for succinct, ambient information design with the level of interaction your views must support. For example, if a view contains touchable subviews, such as buttons, try to stay within an acceptable margin of accuracy for those controls. Shrinking buttons down to fit on a screen can lead to controls that are difficult to touch without concentration. For designers accustomed to desktop and web application development, considering the size of people's fingertips can be a minor obstacle. A good approach is to shift important controls toward the top of your view, where they're easily found and won't be overlapped by hands.

In many cases, subordinate views in your navigation hierarchy will contain more information than will fit on a single screen. You should always design according to your own vision while keeping usability in mind. A good exercise for modeling your navigation might be developing user personae and use cases that focus on the motives of users and the contexts in which they will interact with your application. You'll encounter this advice in many places in this book, as user-centered design is the foundation of successful UX.

Application Templates

The primary tool for developing Cocoa Touch applications is Xcode. Projects created with Xcode typically take advantage of one of the many application templates that Xcode provides. An application template is a built-in starting point for a project and includes a folder structure, class files, and settings that let developers quickly start writing custom code.

The application templates that work best for productivity application development are the navigation-based application template and the tab bar application template. These application templates and their architectural considerations are explored in the next chapter.

Light Utilities

Utility applications differ from productivity applications in that the information they display tends to represent a very simple graph—often a single content object—typically in an ambient design requiring very little interaction. The mobile HIG uses the Weather application as an example of a utility application.

The mobile HIG suggests that utilities should require minimal user input and instead display "glanceable" information on a single view with limited interaction. I agree with the definition, but I think it can be interpreted too narrowly. A slightly different type of iPhone utility is the built-in Calculator application.

Upon startup, the Calculator application is immediately ready for input. The purpose of the application is quite clear. Though the interface is polished, the focus is on usability and clarity. In fact, Calculator lacks the editing controls and "Info" button many utilities provide. This simplicity, power, and usability make the Calculator another excellent example of a streamlined iPhone utility.

When developing utilities, your focus should be on solving a need in a simple way. This goal may be different from solving a simple need. More than any other type of application, utilities need to adhere to the concept of *cooperative single-tasking* outlined in Chapter 5. That is, they should start immediately, shut down quickly, and work with other applications. A utility application should assist the user in the context of a larger task and should be as minimally disruptive to thought processes as possible.

Application Templates

The iPhone application templates that work best for utility development are the utility application template, the view-based application template, and the tab bar application template.

Immersive Applications

Both productivity applications and utilities focus on presenting information to users in a predictable, easily navigable, touch-focused way. Though graphic design can vary wildly among these types of applications, the UI tends to center around standard UIKit controls, such as lists, buttons, text input fields, and search controls. These applications remove all barriers to entry and encourage users to launch, quickly use, and close the applications as often as needed.

Another important type of Cocoa Touch application is the immersive application. You can think of 2D and 3D games, accelerometer-controlled apps, movie players, and the camera as immersive applications. Generally speaking, an immersive app is one that eschews many of the standard controls in favor of a deeper user experience.

It would be reasonable to assume that users anticipate longer launch processes, unconventional controls, and steeper learning curves with immersive applications than with utilities or productivity applications. Popular 3D games often leverage hardware features such as the accelerometer to provide proprietary user experiences. Applications that bring the hardware itself into the foreground are often immersive applications. For example, in the mobile HIG, Apple refers to a sample application that ships with the iPhone SDK called BubbleLevel. The application allows the iPhone or iPod Touch to act as a hardware leveling device with a simulated "bubble level" interface.

It may be tempting to shift all development into immersive applications in the hopes of breaking new ground, avoiding certain constraints in UIKit classes, or more easily porting applications from other technologies. My recommendation is to consider the entire user experience, including the context of its use. Very often, your gut reaction as an experienced UX developer will be correct, but you will still want to keep in mind such concepts as cooperative single-tasking and progressive enhancement.

Application Templates

The iPhone application templates that work best for immersive application development are the OpenGL ES application template, the tab bar application template, the utility application template, and the view-based application template.

CHAPTER 4

Choosing an Application Template

Chapter 3 explored the coarse-grained application types defined by Apple in the mobile HIG. Choosing one of these types for your application is a helpful step in streamlining your UX decision making. The audience personae you define will differ in key ways for each application type. Utility applications are meant to be fast, lightweight, and mostly ambient with flat learning curves. Productivity applications tend to mimic or enhance physical processes, such as writing letters, planning an itinerary, or exploring complex datasets. Because productivity apps exist in a known problem domain, learning curves are often minimal if you employ good design. Immersive applications have steeper learning curves, require more of an investment (time, energy, concentration) for use, and tend to reward users with deeper experiences. Good examples of immersive applications are 3D games and the onboard movie player service.

Still, these three application types are somewhat arbitrary, and are occasionally intermixed to create experiences with multiple modes of interaction. Thinking about the primary bucket in which you would toss your application can help you with another important decision: the application architecture.

If you consider iterative, user-centered design processes for a moment, a project lifecycle might loosely adhere to the following sequence:

1. Identify a problem for a given audience.
2. Develop the core approach for solving the problem.
3. Explore the experience levels, expectations, priorities, and goals of the audience.
4. Develop an initial interface representing the simplest, most familiar implementation.
5. Expose this prototype to representative users and gather feedback.
6. Implement enhancements based on the feedback.
7. Repeat Steps 5 and 6 until a version of the application is ready for release.

To yield the greatest benefits from this process, you'll want to have a relatively stable idea of what your final interface will look like by Step 4. Once you know, roughly, the type of interface you want, you can use Xcode to create your project. Luckily, Xcode provides several core application templates that bootstrap the most common design patterns.

 This book assumes that users are familiar with Xcode and the Objective-C language, perhaps from developing Cocoa applications on the Mac.

To access the New Project window, open Xcode and navigate to File → New Project, or press Apple + Shift + N. Figure 4-1 shows the template chooser in the New Project dialog.

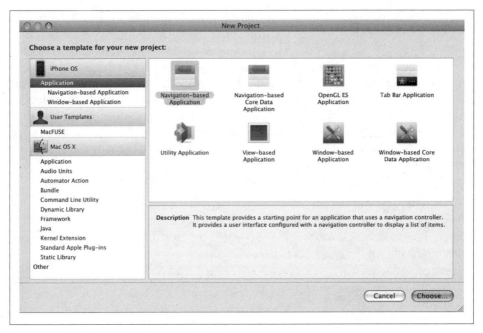

Figure 4-1. New Project dialog in Xcode

Project templates are grouped by platform, such as the iPhone OS or Mac OS X. You can click each template icon to access a very light description of the template. The information is only mildly useful, but it does provide a decent starting point and helpful refresher for each project type. Choosing a template is as simple as selecting it and clicking the Choose button.

Each template generates a project skeleton you can compile and run as a boilerplate application. Naturally, without custom code, the build will be pointless; still, creating one of each, building it, and running the resulting application in the iPhone simulator is a great way to familiarize yourself with each template, including the default executable.

Depending on your vision of how users will interact with your application, you might use one of the default structures and standard user interfaces. After all, each template is quite capable of providing a satisfactory (if generic) user experience.

More likely, however, you will develop custom views that handle their own layouts and draw their own art, extending and enhancing core UIKit classes. It is not uncommon to combine some of the application structures in subordinate fashion. For example, using multiple UITableViewController subclasses in a hierarchy and having UITableViewController subclasses as the view controllers for a tab in a tab-based application template are both popular design patterns.

Apple provides navigation patterns that are quite flexible and will likely work for all but the most creative and distinct immersive applications.

As Apple advises in the mobile HIG, you should prioritize the needs of users and develop the simplest version of an application that reflects your own philosophy and approach to problem solving. This might be an incredibly innovative OpenGL ES application with entirely new modes of interaction, or it might leverage the default designs Apple gives developers. Either way, the sooner you can put prototypes in front of representative users, the sooner you can refine, streamline, incorporate feedback, and—most importantly—ship!

View Controllers

A thorough understanding of the standard interaction design patterns included in UIKit must include view controllers. Cocoa and Cocoa Touch applications tend to follow a classic model-view-controller (MVC) architectural pattern. Domain objects, or *models*, are separated from the *controller* code operating on them. Further separated are the user interface *views* that allow users to interact with controls, input data, and understand data in meaningful ways.

In Cocoa Touch, controllers are most often instances of a *view controller*.

View controllers, as represented in UIKit by the UIViewController class, manage the views that comprise a given Cocoa Touch user interface. Controller classes mediate the communication between models and views, manage asynchronous operations such as networking and threading, and handle the configuration and logic that make your particular application more than the sum of its logical domain objects. In Cocoa Touch, view controllers take on a subset of the application duties by managing the logic specifically around views, with each view controller typically in charge of a single view.

Figure 4-2 shows the relationship for the simplest case: a UIViewController and a UIView.

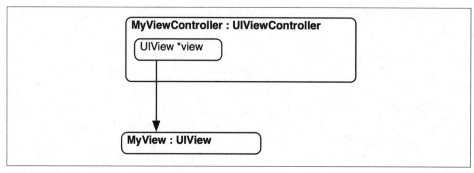

Figure 4-2. A single UIViewController and its UIView

View Controller Subclasses and Corresponding Application Templates

You may wonder why the UIViewController class is being discussed in the context of choosing an application template. The reason is that each application template corresponds to one of several UIViewController subclasses. Choosing the application template that works best for the current iteration of your application requires a basic understanding of the stock view controller subclasses.

There are three types of view controllers you will encounter when creating new iPhone projects. The first is the simplest: the UIViewController. The next is the UITableView Controller. The final view controller subclass is the UITabBarController.

UIViewController and view-based applications

Sometimes the best solution to a problem is a very simple application. For most of us, this is a dream scenario. Xcode supplies an application template that sets up a single view controller and an associated view.

View controller instances, as represented by the UIViewController base class, each have an instance variable called view that points to a UIView or UIView subclass. As mentioned in the previous section, the job of a view controller is to manage all logic for a view, often acting as a delegate for event handling or as a data source.

The view-based application template is a great place to start if your application won't allow users to navigate across a set of data or across a linear process, such as a wizard. In those cases, a navigation-based application template is a more suitable foundation.

If you'd like to use a view-based application template for your application but would prefer to support an application configuration screen in your application, you should consider the utility application template.

UIViewController and utility applications

The utility application template builds upon the simple view-based application template by generating code and Interface Builder files that let users alternate between your primary view and an alternate view. This structure originated in Mac OS X desktop widgets, and is in use for standard, lightweight applications that ship with the iPhone and iPod Touch, such as the Weather application.

UITabBarController and tab-based applications

Tab-based applications give users a very accessible and understandable mechanism for selecting among two or more sibling view controllers. Think of tab controllers the way you might think of TV channels. You can touch a tab to switch to its controller and view, which together make up the content for that tab. Proper use of tabs is to show separate functional areas of an application as separate but equal tabs. In other words, you don't want to have a high-level tab, and then a tab next to it that represents a lower-level bit of functionality that might otherwise belong "under" the first. The Tweetie application by Atebits uses a tab bar to display various types of messages on Twitter. For example, there is a tab for your main Twitter feed, one for replies sent to you, one for direct messages, and one for favorites.

Figure 4-3 shows the result of compiling the default tab bar application template.

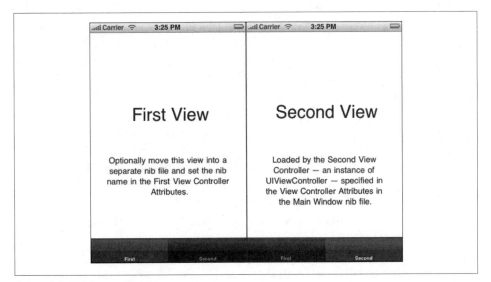

Figure 4-3. Tab bar application

Note the two tabs at the bottom of each screen. Technically, selecting a tab switches the current `UIViewController` pointed to by the main `UITabBarController` instance. From a non-technical user perspective, selecting the second tab declares to the application that the user would like to focus on a different area of functionality from that in the first tab.

Each tab bar item in the tab bar represents a single `UIViewController` or `UIViewController` subclass. That means you can choose to combine your view controllers, such as using a `UINavigationController` or `UITableViewController` "inside" one of your tabs.

 One note of advice: retrofitting an application into a tab-based application can be somewhat tedious, though far from the end of the world. Still, if you think you might switch to a tab-based architecture in the near future, it's worth using the tab bar application template from the beginning of your development.

UINavigationController and navigation-based applications

A navigation controller is a special `UIViewController` subclass that allows users to build and drill into a stack of `UIViewController` instances. This is useful when users traverse nodes in a tree or linked list data structure.

Think of the stack of view controllers as a stack of paper. The navigation controller manages the stack and lets users trigger actions that add a new sheet to the top of the stack or that, alternately, remove a sheet from the stack. To programmers, this is called "pushing" and "popping" items onto a stack.

When you tell a navigation controller to push a new view controller and view onto the top of the stack, the navigation controller will automatically load the new controller and view into memory and then trigger an animation that slides the current view off-screen, to the left, while sliding the new view onscreen, from the right. You can see this in the Mail application when selecting a message to read, for example. This mechanism is seen consistently among a large number of Cocoa Touch applications and has, in some ways, become an iconic transition for the platform.

Figure 4-4 shows the relationship between a `UINavigationController` and its subordinate `UIViewController` instances.

When prototyping your application, try to decide whether you will be navigating down through a hierarchical dataset, such as a list of messages. If so, consider choosing an application template that builds upon a `UINavigationController` foundation.

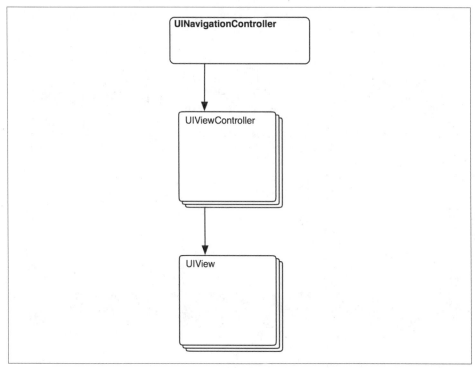

Figure 4-4. UINavigationController and UIViewController relationships

UITableViewController and navigation-based applications

Table views are used to manage a list of table cells, with each cell representing a distinct item of some sort. Typically, there is a one-to-one relationship among onscreen table view cells and model objects. For example, the Contacts application that ships with the iPhone maps onscreen cells, or rows, with each user stored in the address book.

When using a `UITableViewController` in a navigation-based application, the typical pattern is to represent data as a tree with the first level of objects listed in the table. When tapping a single cell, a new view controller is added to the top of the navigation controller stack and its view is transitioned in from right to left. Depending on your data structure, the new view controller might be another `UITableViewController` representing all child nodes for the selected node, or it might be a custom `UIView` that shows custom art and layout.

Given the extensibility of `UITableViewControllerCell` instances, it's worth exploring the use of table views for many problems.

Figure 4-5 shows screenshots of an application that traverses a set of nodes using sequential `UITableViewControllers` on a `UINavigationController` stack.

Figure 4-5. A series of UITableViewControllers in a UINavigationController stack

OpenGL ES applications

Applications that are based on the OpenGL ES application template do not use `UIViewController` subclasses. Instead, the default template uses an `EAGLView`—a subclass of `UIView`—to an instance variable in the application delegate class. The `EAGL View` class sets up an OpenGL ES context, using an instance of the `EAGLContext` class. The `EAGLView` instance also creates a timer that triggers a routine for drawing to the `EAGLContext`.

This book doesn't cover OpenGL ES applications in depth, but generally you will want to use OpenGL ES to develop any 3D games. Additionally, many 2D immersive applications could take advantage of the power of OpenGL ES to do custom drawing.

Core Data Templates

The iPhone 3.0 update to the iPhone OS and iPhone SDK included a great new feature, ported from Mac OS X, called Core Data. Core Data is used to store objects in a local database that can be queried later on. This is a great alternative to using "raw" SQLite on iPhone 2.0 and earlier (though SQLite remains an excellent option).

You will notice that the project template window in Xcode includes options for navigation-based and window-based Core Data applications. These options are exactly like their non-Core Data templates with a few extra classes, protocols, and methods for managing Core Data. The decision to use Core Data versus SQLite for state management and information storage is not really a question of user experience, and there are many factors that might influence your choice.

If you are unsure whether to use Core Data while prototyping an application, it may be best to go ahead and choose the Core Data version of the appropriate template. Adding Core Data to an existing project is not a terrible process, but for prototyping purposes, it may be more productive to include it at the beginning and remove it later if it proves unnecessary.

CHAPTER 5

Cooperative Single-Tasking

Lots of veteran Mac programmers like to joke about the trials and tribulations of cooperative multitasking as it existed on the Mac OS prior to OS X. For those unfamiliar with the concept, cooperative multitasking is a sort of time-sharing pattern for computer processes. In a cooperative multitasking system, a selected process will take focus and use the majority of available resources to perform its work. After a time, that process should yield control of those resources to other, waiting processes. In this model, however, it's not uncommon for a bug in one application to "freeze" the entire operating system by blocking access to the CPU. This was a common enough problem in pre-OS X days that the power button on older Macs often had as much wear and tear as the space bar on the keyboard.

The iPhone OS leverages an operating system that, at its core, handles multitasking very well. Unfortunately, Apple has given developers of iPhone applications a set of rules that prevent access to true multitasking. We know that, for all intents and purposes, custom Cocoa Touch applications are not allowed to run in the background. This means that when an application isn't in focus and doesn't have full control of the screen, code from that application cannot run.

There are obvious exceptions to this rule. For example, the Phone, iPod, Mail, SMS, and Alarm applications take advantage of background processes to perform their functions. Unless a special relationship is worked out with Apple, though, no third-party application is allowed to operate in the background.

This is, in some ways, similar to the concept of cooperative multitasking. For example, with cooperative multitasking, users feel as though they have a single application in the foreground, although they may have many applications open in a dimmed sort of "paused" state. The foreground application has the bulk of available resources within its domain.

Task Management and iPhone OS

The implementation of application focus for Cocoa Touch is not based on cooperative multitasking. Apple provides for external interruptions and terminations of apps, and although Apple gives developers no access to multiprocess programming, the operating system is designed for it.

The term *cooperative single-tasking* is a way of describing the following rule set:

- Your application should start quickly.
- Your application should deactivate (or pause) elegantly and responsibly.
- Your application should reactivate (or unpause) elegantly.
- Your application should terminate quickly when asked.
- Your application should leverage custom URLs to cooperate with other applications.
- When possible, your application should use shared data sources, such as the Address Book, to share common data between applications.
- If your application is part of a suite of applications, you should attempt to share preferences where overlap occurs.

Example Application

The example in this section is a small sample application called ImageSearch that searches Google Images for a user-provided term or phrase. The application launches quickly, handles interruptions, handles terminations, and responds to custom URLs. In addition, the application attempts to save state appropriately for all common exit points, and restores that state incrementally when relaunched. While it doesn't illustrate all the concepts of cooperative single-tasking, the application focuses on becoming a friendly part of the iPhone OS and of anticipated user experience.

The ImageSearch application is created for users who want to search Google Images in the quickest way possible. The user may also want to save past search terms and results.

This application was created with the view-based controller template in Xcode:

```
// ImageSearchViewController.h
#import <UIKit/UIKit.h>
#import <QuartzCore/QuartzCore.h>

@interface ImageSearchViewController : UIViewController <UIWebViewDelegate,
UISearchBarDelegate> {
    UIWebView *webView;
    UIImageView *lastKnownView;
    NSMutableDictionary *lastKnownSearch;
    BOOL lastKnownSearchIsDirty;
    NSString *searchTermFromURL;
}

@property (nonatomic, assign) NSString *searchTermFromURL;

- (void) loadLastKnownSearch;
- (void) performSearchWithTerm:(NSString *)searchTerm;
- (void) performSearchForSearchBar:(UISearchBar *)theSearchBar;
```

```
- (void) createLastKnownSearchSnapshot;
- (void) prepareForTermination;
- (void) setLatestURLString:(NSString *)theURLString;
- (NSString *) latestURLString;
- (void) setLatestSearchTerm:(NSString *)theTerm;
- (NSString *) latestSearchTerm;
- (void) reloadLastKnownSearch;
- (void) loadLastKnownSearchImageFromCache;

@end
```

The implementation file shows all logic for the `ImageViewSearchController` class. The remainder of this chapter will refer to this class to explain the programming model used in implementing both launch and exit strategies:

```
// ImageSearchViewController.m
#import "ImageSearchViewController.h"

static NSString * const kSearchURLString =
@"http://www.google.com/m/search?q=%@&site=images";
#define CONTENT_WIDTH 320.0
#define CONTENT_HEIGHT 434.0
#define SEARCH_BAR_HEIGHT 44.0
#define CONTENT_Y 46.0

@implementation ImageSearchViewController

@synthesize searchTermFromURL;

#pragma mark UIView loading methods
- (void)loadView
{
    //    Create our main view.
    UIView *view = [[UIView alloc] initWithFrame:[[UIScreen mainScreen]
applicationFrame]];

    //    Set the autoresizing mask bits to allow flexible resizing if needed.
    view.autoresizingMask =
        UIViewAutoresizingFlexibleHeight|UIViewAutoresizingFlexibleWidth;

    //    Create the search bar.
    CGRect searchBarFrame = CGRectMake(0.0, 0.0, CONTENT_WIDTH,
        SEARCH_BAR_HEIGHT);
    UISearchBar *searchBar = [[UISearchBar alloc]
        initWithFrame:searchBarFrame];
    searchBar.delegate = self;
    [view addSubview:searchBar];

    //    Assign the UIView to our view property.
    self.view = view;
    [view release];
}

- (void)viewDidLoad
{
```

```objc
    //    Let the call stack close so the Default.png file will disappear.
    [super viewDidLoad];

    //    Shift the time-intensive load off to a new call stack.
    //    You can also extend this to spin off a new thread, which would
    //    allow users to interact with any already present UI.
    if(searchTermFromURL == nil){
        [self performSelector:@selector(loadLastKnownSearch)
            withObject:nil afterDelay:0.01];
    }else{
        [self performSelector:@selector(performSearchWithTerm:)
            withObject:searchTermFromURL afterDelay:0.01];
    }
}


- (void) buildWebView
{
    CGRect frame = self.view.frame;
    frame.origin.y = CONTENT_Y;
    frame.size.height = CONTENT_HEIGHT;
    webView = [[UIWebView alloc] initWithFrame:frame];
    webView.autoresizingMask =
        UIViewAutoresizingFlexibleHeight|UIViewAutoresizingFlexibleWidth;
    webView.delegate = self;
}


#pragma mark Search methods
- (void) loadLastKnownSearch
{
    NSUserDefaults *defaults = [NSUserDefaults standardUserDefaults];
    lastKnownSearch = (NSMutableDictionary *)[defaults
        dictionaryForKey:@"lastKnownSearch"];

    if(lastKnownSearch == nil){
        lastKnownSearch = [[NSMutableDictionary alloc] init];
        return;
    }

    [self reloadLastKnownSearch];

    [self loadLastKnownSearchImageFromCache];
}

- (void) loadLastKnownSearchImageFromCache
{
    NSString *lastKnownViewPath =
        [NSString stringWithFormat:@"%@/../Documents/lastKnownView.png",
        [[NSBundle mainBundle] bundlePath]];
    NSFileManager *manager = [NSFileManager defaultManager];
    if([manager fileExistsAtPath:lastKnownViewPath]){
        UIImage *lastKnownViewImage = [UIImage
                                imageWithContentsOfFile:lastKnownViewPath];
        lastKnownView = [[UIImageView alloc] initWithImage:lastKnownViewImage];
        CGRect frame = lastKnownView.frame;
```

```
            frame.origin.y = CONTENT_Y;
            frame.size.height = CONTENT_HEIGHT;
            lastKnownView.frame = frame;
            NSLog(@"adding subview: lastknownview");
            [self.view addSubview:lastKnownView];
    }
}

- (void) performSearchWithTerm:(NSString *)searchTerm
{
    NSString *theURLString = [NSString stringWithFormat:kSearchURLString,
                                searchTerm];

    NSURL *theURL = [NSURL URLWithString:theURLString];
    NSURLRequest *theRequest = [NSURLRequest requestWithURL:theURL];

    if(webView == nil){
        [self buildWebView];
    }

    [webView loadRequest:theRequest];
    [lastKnownSearch setValue:searchTerm forKey:@"searchTerm"];
    [self setLatestURLString:theURLString];
    lastKnownSearchIsDirty = YES;
}


#pragma mark Rehydrating the last known search
- (void) reloadLastKnownSearch
{
    NSURL *theURL = [NSURL URLWithString:[self latestURLString]];
    NSURLRequest *theRequest = [NSURLRequest requestWithURL:theURL];

    if(webView == nil){
        [self buildWebView];
    }

    [webView loadRequest:theRequest];
    lastKnownSearchIsDirty = YES;
}


#pragma mark Managing the "history"
- (void) setLatestSearchTerm:(NSString *)theTerm
{
    [lastKnownSearch setValue:theTerm forKey:@"searchTerm"];
}

- (NSString *) latestSearchTerm
{
    return [lastKnownSearch valueForKey:@"searchTerm"];
}

- (void) setLatestURLString:(NSString *)theURLString
{
```

```
        [lastKnownSearch setValue:theURLString forKey:@"theURLString"];
    }

    - (NSString *) latestURLString
    {
        return [lastKnownSearch valueForKey:@"theURLString"];
    }

    - (void) clearCachedSearch
    {
        NSLog(@"clearCachedSearch finds subviews: %@", self.view.subviews);
        [lastKnownView removeFromSuperview];
        [self.view setNeedsDisplay];
    }

#pragma mark UISearchBarDelegate methods
    - (void) searchBarTextDidBeginEditing:(UISearchBar *)searchBar
    {
        [self clearCachedSearch];
    }

    - (void) searchBarTextDidEndEditing:(UISearchBar *)searchBar
    {
        [self performSearchForSearchBar:searchBar];
    }

    - (void) searchBarSearchButtonClicked:(UISearchBar *)searchBar
    {
        [self performSearchForSearchBar:searchBar];
    }

    - (void) performSearchForSearchBar:(UISearchBar *)theSearchBar
    {
        NSString *newSearchTerm = [theSearchBar text];
        if(newSearchTerm == nil){
            return;
        }
        [self performSearchWithTerm:newSearchTerm];
    }


#pragma mark UIWebViewDelegate methods
    - (void)webViewDidFinishLoad:(UIWebView *)theWebView
    {
        NSString *loc = [[webView.request URL] absoluteString];
        [self setLatestURLString:loc];
        [self.view addSubview:webView];
        [lastKnownView removeFromSuperview];
    }


#pragma mark Termination methods
    - (void) prepareForTermination
    {
        if(lastKnownSearchIsDirty){
```

```
        [self createLastKnownSearchSnapshot];
    }
}

- (void) createLastKnownSearchSnapshot
{
    CGRect rect = webView.frame;

    //    Scroll to the top for the screencap.
    [webView stringByEvaluatingJavaScriptFromString:@"window.scrollTo(0, 0);"];

    UIGraphicsBeginImageContext(rect.size);
    CGContextRef currentContext = UIGraphicsGetCurrentContext();
    CALayer *contentsLayer = webView.layer;
    [contentsLayer renderInContext:currentContext];

    UIImage *image = UIGraphicsGetImageFromCurrentImageContext();

    //    Close this transaction
    UIGraphicsEndImageContext();
    NSString *path =
        [NSString stringWithFormat:@"%@/../Documents/lastKnownView.png",
          [[NSBundle mainBundle] bundlePath]];
    NSData *d = UIImagePNGRepresentation(image);
    [d writeToFile:path atomically:NO];

    //    Save the strings for the search.
    NSUserDefaults *defaults = [NSUserDefaults standardUserDefaults];
    [defaults setObject:lastKnownSearch forKey:@"lastKnownSearch"];
    lastKnownSearchIsDirty = NO;
}

- (void)dealloc
{
    [searchTermFromURL release];
    [webView release];
    [lastKnownView release];
    [lastKnownSearch release];
    [super dealloc];
}

@end
```

Launching Quickly

Consider the most likely interaction pattern for an iPhone: the user pulls the device out of his pocket, taps a few times to accomplish a fixed task, then locks the device and puts it back in its place.

Now consider the most desirable user experience: the user is completing a task without focusing fully on their mobile device. Think about a user walking down the street, or on the subway, using an iPhone application but not able to give total focus to it because of the environment.

If launching your application shows an introductory splash image with your amazing company logo for 5 to 10 seconds before becoming useful, you will have removed the fluid, continuous flow from an otherwise smooth user experience. Every moment spent waiting to accomplish a distinct task is time spent considering your competition. Instead of splash screens, it's worth looking at the fastest way to accomplish two goals:

- Prepare the application for user input or display of information.
- Give the illusion that your application was always open, waiting in the wings, and that it will be there next time, ready to go.

Use cases should accomplish both goals while following this sequence:

1. The user touches the application icon.
2. The structure of the application user interface loads.
3. The data (text, images) load "into" the user interface.

To accomplish the first goal, Apple suggests shipping a 320 × 480 pixel portable network graphics (PNG) file with your application. The file should act as a graphical stand-in for your application views. That is, you should design the image so that it looks as close to the initial state of your application as possible. For a standard navigation-based application that presents information via a `UITableView`, your launch image would most likely look like a table without data. When the data is ready, the real UI automatically replaces the launch image.

You can smooth the steps by including in the image any immutable data that, though dynamic by nature, will remain the same inside your application. For example, if the root view controller always has the same title, you can safely include that text in your launch image.

To leverage the built-in launch image functionality, include the PNG file at the top level of your application bundle. The image should be named according to the following criteria:

- If the launch image is globally useful (that is, not internationalized or localized), name it *Default.png* and place it at the top level of your application bundle.
- If the launch image contains text that should be internationalized or localized, name it *Default.png* but place it into the appropriate language-specific bundle subdirectory.
- If the launch image represents a specific URL scheme that would lead to your application launching with a non-default view, name the file *Default-<scheme>.png*, where *scheme* is replaced with the URL scheme. For example, the Mail app might display a launch image called *Default-mailto.png* when handling a `mailto://` link.

All of this is to give the impression that users aren't so much launching and quitting applications, but merely switching between them. Consider the familiar hot-key pattern in OS X on the desktop: when users combine the Command and Tab keys, they can switch between open applications. This is more desirable than the feeling of launching and quitting applications, and it reinforces the fluidity and continuity of the iPhone OS.

When considering a splash screen for your launch images, imagine how it would feel on a Mac OS X desktop machine to have to view a splash screen every time you switched between applications.

Example Application

The ImageSearch example uses a lightweight, informative launch image that draws the standard interface along with a callout box with instructions. This could easily be localized by setting the appropriate *Default.png* file for each localized bundle.

The launch process completes a minimum amount of work as efficiently as possible in order to present the illusion of continuity and state persistence, completing as quickly as possible before handing control over to the application.

The launch image is visible during the launch process until the first screen of the application—which is also as lightweight as possible—is loaded.

Next, the main screen loads in a partially dynamic state. A usable search bar is instantiated so a user can begin entering a search term. This occurs during the loadView: method of the ImageSearchViewController class, which is used to build the customized UIView instance assigned to the view property:

```
- (void)loadView
{
    // Create our main view.
    UIView *view = [[UIView alloc] initWithFrame:[[UIScreen mainScreen]
        applicationFrame]];

    // Set the autoresizing mask bits to allow flexible resizing if needed.
    view.autoresizingMask =
        UIViewAutoresizingFlexibleHeight|UIViewAutoresizingFlexibleWidth;

    // Create the search bar.
    CGRect searchBarFrame = CGRectMake(0.0,
        0.0,
        CONTENT_WIDTH,
        SEARCH_BAR_HEIGHT
    );

    UISearchBar *searchBar = [[UISearchBar alloc] initWithFrame:searchBarFrame];
    searchBar.delegate = self;
    [view addSubview:searchBar];

    // Assign the UIView to our view property.
    self.view = view;
    [view release];
}
```

The `viewDidLoad:` method breaks out of the main call stack with a call to `performSelec`
`tor:withObject:afterDelay:` to trigger the invocation of `loadLastKnownSearch:` after a
trivial delay of 0.01 seconds. This enables the `viewDidLoad:` method to return more
quickly, decreasing the perceived load time:

```
- (void)viewDidLoad
{
    // Let the call stack close so the Default.png file will disappear.
    [super viewDidLoad];

    // Shift the time-intensive load off to a new call stack.
    // You can also extend this to spin off a new thread, which would
    // allow users to interact with any already present UI.
    if(searchTermFromURL == nil){
        [self performSelector:@selector(loadLastKnownSearch) withObject:nil
            afterDelay:0.01];
    }else{
        [self performSelector:@selector(performSearchWithTerm:)
            withObject:searchTermFromURL afterDelay:0.01];
    }
}
```

If a search has been executed in a prior session, the area below the search bar is dedi-
cated to a dimmed, cached bitmap snapshot loaded from the *Documents* directory. The
bitmap snapshot is created, if needed, in the `applicationWillTerminate:` method when
the application exits. The loading of the cached bitmap occurs in the `loadLastKnown`
`Search:` method. In parallel, the example begins loading the search results from the
previous session. If no saved search exists, the method simply exits:

```
- (void) loadLastKnownSearch
{
    NSUserDefaults *defaults = [NSUserDefaults standardUserDefaults];
    lastKnownSearch =
        (NSMutableDictionary *)[defaults
            dictionaryForKey:@"lastKnownSearch"];

    if(lastKnownSearch == nil){
        lastKnownSearch = [[NSMutableDictionary alloc] init];
        return;
    }

    [self reloadLastKnownSearch];

    [self loadLastKnownSearchImageFromCache];
}
```

If a user begins a new search by typing into the search bar before the prior search has
reloaded in the `UIWebView`, the cached representation is removed in anticipation of a
new query. This is accomplished using the `searchBarTextDidBeginEditing:` delegate
method from the `UISearchBarDelegate` interface:

```
- (void) searchBarTextDidBeginEditing:(UISearchBar *)searchBar
{
    [self clearCachedSearch];
}
```

The process of launching the application and completing a search involves the following sequence:

1. The user taps an application icon.

2. The launch image is loaded automatically by the OS and is animated to full screen as the *face* of the application while it loads.

3. The controller builds a very, very lightweight initial interface that supplies the search field for task-oriented users. Below it, if it exists, is a reference to the last known search results screen. This lets users double-check older results and improves the feeling of continuity.

4. Unless a user begins typing in a new query, the query starts to invisibly reload over HTTP.

5. When the expensive, asynchronous HTTP request completes, the application displays the result in place of the cached view. Undimming the user interface subtly alerts the user that interaction is enabled.

If the design of the application did not focus on providing for continuity, a simpler but less friendly flow would be to simply reload the last query while showing a splash screen of some sort, effectively blocking interaction for any number of seconds and frustrating the user.

Handling Interruptions

There are two occasions when your application might need to give up focus. The first is when a dialog overlay is triggered externally, such as with an incoming SMS message or phone call. The second is when your application is running but the user clicks the lock button or the phone triggers its autolock mechanism in response to a lack of user input.

The most important thing for application developers to remember when considering interruptions is that, because your application is being interrupted somehow, a user may take an action that causes the termination of your application. A good example is when a phone call comes in and triggers the familiar "answer or ignore" dialog. If the user ignores the call, your application will regain focus. If the user answers the phone, your application will go into a suspended state.

The UIApplicationDelegate protocol that defines an application delegate includes three methods that handle interruptions:

- The applicationWillResignActive: method is called when the OS decides to shift your application out of the primary focus. Any time this method is invoked, you might want to look for operations that can be paused to reflect the lower (visual) priority state of your application. For example, if you are doing heavy graphics processing for display, or using resource-hungry features such as the accelerometer for user input, you can likely pause those operations.

- The method `applicationDidBecomeActive:` is called when your application takes the coveted primary focus of the iPhone OS. In this method, you should check for any paused processes or switched flags and act appropriately to restore the state users expect.

- Finally, the `applicationWillTerminate:` method is called when the iPhone OS tells the application to exit. This happens often in response to memory management problems, such as memory leaks, and when inter-application communication occurs. For example, invoking a new email message in the Mail application via a `mailto://` link would tell the OS first to terminate your application and then to launch the Mail app. To users, of course, this should feel as close to simple application switching as possible.

It is vital that an application terminate as quickly as possible, while maintaining state for its next invocation. The section "Handling Terminations" on page 51 covers the termination of your application.

Interruptions and the Status Bar

When a user opens your application while on a call, the status bar at the top of the screen will grow taller and contain messaging to remind users that they are on a call. Additionally, the status bar will act as a shortcut to close your application and return to the main Phone application screen for an inbound call. You should test your application for this scenario to ensure that your view and all its subviews are laying themselves out properly to reflect the change in available real estate.

Example Application

The ImageSearch application does very little intensive processing. It's a simple application that acts as a semi-persistent search tool for Google Images.

For clarity, the application delegate class, `ImageSearchAppDelegate`, includes a flag for storing whether the application is in the foreground. The flag is a `BOOL` called `isForegroundApplication`, and flag can be used to determine whether the application is in active or inactive mode when termination occurs. In more complex applications, the termination process may have different requirements and cleanup needs:

```
// ImageSearchAppDelegate.h
#import <UIKit/UIKit.h>

@class ImageSearchViewController;

@interface ImageSearchAppDelegate : NSObject <UIApplicationDelegate> {
```

```
    UIWindow *window;
    ImageSearchViewController *viewController;
    BOOL isForegroundApplication;
}

@property (nonatomic, retain) IBOutlet UIWindow *window;
@property (nonatomic, retain) IBOutlet ImageSearchViewController *viewController;

- (void) performSearch:(NSString *)searchTerm;

@end
```

The implementation file shows example log messages that print messages based on the isForegroundApplication flag:

```
#import "ImageSearchAppDelegate.h"
#import "ImageSearchViewController.h"

@implementation ImageSearchAppDelegate

@synthesize window;
@synthesize viewController;


- (void) applicationDidFinishLaunching:(UIApplication *)application
{
    NSLog(@"applicationDidFinishLaunching.");
    isForegroundApplication = YES;
    // Override point for customization after app launch
    [window addSubview:viewController.view];
    [window makeKeyAndVisible];
}

- (void) applicationWillTerminate:(UIApplication *)application
{
    NSLog(@"Ok, beginning termination.");
    if(isForegroundApplication){
        NSLog(@"Home button pressed or memory warning. Save state and bail.");
    }else{
        NSLog(@"Moved to the background at some point. Save state and bail.");
    }
    [viewController prepareForTermination];
    NSLog(@"Ok, terminating. Bye bye.");
}

- (void) applicationWillResignActive:(UIApplication *)application
{
    NSLog(@"Moving to the background");
    isForegroundApplication = NO;
}

- (void) applicationDidBecomeActive:(UIApplication *)application
{
    NSLog(@"Moving from background to foreground.");
    isForegroundApplication = YES;
}
```

```
#pragma mark Custom URL handler methods
/*
    The URI structure is:
    imagesearch:///search?query=my%20term%20here
*/
- (BOOL)application:(UIApplication *)application handleOpenURL:(NSURL *)url
{
    BOOL success = NO;
    if(!url){
        return NO;
    }

    // Split the incoming search query into smaller components
    if ([@"/search" isEqualToString:[url path]]){
        NSArray *queryComponents = [[url query] componentsSeparatedByString:@"&"];
        NSString *queryComponent;
        for(queryComponent in queryComponents){
            NSArray    *query = [queryComponent componentsSeparatedByString:@"="];
            if([query count] == 2){
                NSString *key = [query objectAtIndex:0];
                NSString *value = [query objectAtIndex:1];

                if ([@"query" isEqualToString:key]){
                    NSString *searchTerm =
                        (NSString *)CFURLCreateStringByReplacingPercentEscapes(
                            kCFAllocatorDefault,
                            (CFStringRef)value, CFSTR("")
                        );
                    [self performSearch:searchTerm];
                    [searchTerm release];
                    success = YES;
                }
            }
        }
    }
    return success;
}

- (void) performSearch:(NSString *)searchTerm
{
    viewController.searchTermFromURL = searchTerm;
    [viewController performSearchWithTerm:searchTerm];
}

- (void) dealloc
{
    [viewController release];
    [window release];
    [super dealloc];
}

@end
```

Handling Terminations

Developers should be careful to spend as much time focusing on their exit processes as on their launch processes. The ideal cooperative model, when applied to an application, would result in a very symmetrical curve where launch and termination are short and engagement is immediate.

The way in which you terminate your application will probably depend on the state of the application and its resources. Obvious goals such as committing all pending transactions and freeing used memory are important, but what about preparing for the next launch? Should you cache the whole screen? If so, should you go beyond the clipped area that is visually present in order to provide for a scrollable placeholder in the future? What will your users expect? Is the nature of your application such that state persistence matters? Will the criteria depend on the frequency between launches? What if the user waits a month to relaunch your application? What if the wait is five seconds?

The overarching goal is to make opening and closing as painless as possible and to follow good UX and anticipate the expectations of your users.

Here are some guidelines for streamlining terminations:

- Perform as few IO tasks as possible. Saving locally will be safer than saving over a network.
- If saving over a network, provide an internal timeout that invalidates your request and closes the application.
- Don't save all persistence tasks until the termination call. Perform database or filesystem writes when they are relevant instead of caching for one big save. There is a balance to find here, but if you consider event-based writes in your design, you can make informed decisions.
- Stop any non-critical, expensive operations instead of letting them finish. If your main event loop is blocked, your application cannot terminate smoothly.

Example Application

The ImageSearch application hedges against two asynchronous startup processes by caching information as follows:

1. The application loads the main application into memory and creates the default view in the `loadView:` method.
2. The last search is restored using an HTTP request inside a `UIWebView`.

For the first step, the solution takes a snapshot of the `UIWebView` on termination and saves it to the *Documents* directory. This allows it to be reloaded later. A similar caching operation is used by Apple for its applications and would be a welcome addition to the official SDK. Currently, there are undocumented API calls to generate snapshots, but this book prefers to focus on the official SDK. As of now, an event-based caching model and two-step launch process is a strong and simple pattern.

The second step—reloading the last-used HTTP request and effectively returning users to the place they were when they last used the application—is more complex. With all network operations, and especially those on a mobile device, the chance of errors, latency, and user frustration are relatively high. The second step of our two-step launch process mitigates these concerns by displaying a cached representation of the last known search while the live version is loaded.

The application takes a snapshot immediately prior to termination. The snapshot could be generated when the page loads, but doing so would waste resources. Only the last-known page needs to be converted to a snapshot. Additionally, taking a small bitmap snapshot of the screen and dimming it with the CoreGraphics graphics framework is a shockingly trivial operation and fits within the expected termination flow. This is a good example of considering the big picture when optimizing applications for real— as opposed to ideal—usage.

Using Custom URLs

Anyone with experience developing web applications may recognize parts of cooperative single-tasking from the architecture of the Web.

If you compare a common user agent, such as a single tab within Safari, to a running iPhone application, the parallels are interesting:

- Only one main resource URI may be rendered at a time. In other words, only one web page can be viewed in the browser.
- Resources can link to each other and pass data via the HTTP protocol, but they cannot generally speak to each other in a more intimate fashion, such as database-to-database.

Using links, buttons, or events inside one application to launch another application adheres very well to the concept of cooperative single-tasking. By working together instead of competing, applications become parts in a larger, more consistent user experience.

You can use links to open the default iPhone applications, such as Mail, Phone, Safari, and SMS. The following example illustrates a link to the Phone application:

```
[[UIApplication sharedApplication]
 openURL:[NSURL URLWithString:@"tel://18005551212"]];
```

The openURL: call passes an instance of NSURL that points to a resource located at tel:// 18005551212. The protocol handler, tel://, is registered with the operating system by the Phone application. When any URL fitting that scheme is called from any app, the Phone application will open and the calling application will terminate.

You can register your own custom scheme and handle calls from other applications fairly simply on the iPhone OS. The first thing you need to do is register your scheme for the application in the *Info.plist* file. To do so, add a new key to *Info.plist* called CFBundleURLTypes. For the value, replicate the values in the following code, changing the package identifier to reflect your organization, business, or client:

```
<key>CFBundleURLTypes</key>
    <array>
        <dict>
            <key>CFBundleURLName</key>
            <string>com.tobyjoe.${PRODUCT_NAME:identifier}</string>
            <key>CFBundleURLSchemes</key>
            <array>
                <string>imagesearch</string>
            </array>
        </dict>
    </array>
```

Next, define your URL handler method:

```
#pragma mark Custom URL handler methods
/*
 The URI structure is:
 imagesearch:///search?query=my%20term%20here
*/
- (BOOL)application:(UIApplication *)application handleOpenURL:(NSURL *)url
{
    BOOL success = NO;
    if(!url){
        return NO;
    }
    if ([@"/search" isEqualToString:[url path]]){
        NSArray *queryComponents = [[url query] componentsSeparatedByString:@"&"];
        NSString *queryComponent;
        for(queryComponent in queryComponents){
            NSArray     *query = [queryComponent componentsSeparatedByString:@"="];
            if([query count] == 2){
                NSString *key = [query objectAtIndex:0];
                NSString *value = [query objectAtIndex:1];

                if ([@"query" isEqualToString:key]){
                    NSString *searchTerm =
                        (NSString *)CFURLCreateStringByReplacingPercentEscapes(
                            kCFAllocatorDefault,
                            (CFStringRef)value, CFSTR("")
                        );
                    [self performSearch:searchTerm];
                    [searchTerm release];
                    success = YES;
                }
            }
        }
    }
    return success;
}
```

The query term is parsed from the custom URL and is passed to the view controller for execution.

With iPhone OS 3.0 and later, developers can easily detect the presence of a custom URL scheme. The `UIApplication` class in iPhone OS 3.0 and later includes a method called `canOpenURL:` that accepts an `NSURL` and returns a `BOOL` representing the current support on the device for a URL scheme.

The following example detects whether a fictional application, FooBar, is installed by testing the existence of a `foobar://` URL scheme. If the scheme can be handled, it means that the application is present and that button for switching to that application can be safely displayed:

```
- (void)displayFooBarButtonIfFooBarIsInstalled
{
    UIApplication *myApp = [UIApplication sharedApplication];
    NSURL myURL = [NSURL
    URLWithString:@"foobar://doSomethingWith?message=hello+world"];
    if([myApp canOpenURL:myURL]){
        // Show the button for the FooBar app
        myFooBarLaunchButton.hidden = NO;
    }else{
        // Hide the button.
        myFooBarLaunchButton.hidden = YES;
    }
}
```

Prior to iPhone OS 3.0, calling an unknown URL scheme resulted in an unsightly alert box telling users that the scheme is not supported. Craig Hockenberry developed a method to test for the presence of a custom URL scheme while developing Twitterrific. You can study his sample code at his blog via *http://furbo.org/2008/10/01/redacted/*.

Using Shared Data

The iPhone OS creates a sandbox of sorts for each application. *Sandboxes* are restricted spaces reserved for an application. The application cannot be accessed by code running outside the sandbox, nor can it access external code or files. As you have seen, communication between applications happens mostly by way of custom URL schemes, through which small amounts of string-based information may be transferred. In most cases, any files created by an application are visible only to that application. Apple has made exceptions to this for some types of user data. Examples are contacts (as in the Contacts application) and photographs (as in the Photos application).

Apple exposes APIs for accessing contacts at a fairly low level. For photos, access primarily occurs by way of a standard user-facing control that is invoked inside applications.

The example application doesn't use any of the shared data services, but possible upgrades would include allowing a user to select a person from the address book and use that person's name as a search term.

Using Push Notifications

When Apple announced that the iPhone OS would allow only one userland application (i.e., a non-Apple application that the user installs) to run at a time, many developers were upset. Blogs and mailing lists buzzed with conversations in which developers felt they were being locked down too tightly. Apple then presented an alternative to having some applications run in the background. Using current buzzwords, the Apple solution was to have the application run in "the cloud" instead. That is, your application could be developed with an analogous online service that could keep track of external events and, at important points, could notify a service hosted by Apple of the change. Apple could then push the notification to related devices, with the result being that the iPhone OS could add a *numeric badge graphic* to the application icon on the Home screen. The numeric badge is a standard user interface enhancement for Cocoa applications. When the desktop Mail application receives new email messages, it adds a small red circle or oval to the Mail icon in the Dock. Inside the red shape is a number indicating the number of new, unread emails.

 The push service can also display an alert message on the device and play a short sound (using audio files included in an application) if developers wish. Generally, though, it is best to avoid such disruptive notifications, saving them for very important circumstances. The downsides of disruptive alerts are covered later in "Bullhorns" on page 152.

Take for example an instant messaging client. A developer could release a client application that runs natively on the iPhone, but only in the foreground. If, upon termination of the application, the user remains available for messaging within the online instant messaging service, that service could notify Apple when a new message comes in. Apple would then push a notice to a device, leading to a numeric badge on the application icon representing the number of unread messages.

Another example is an online image hosting service such as Flickr. A developer could create an iPhone application that uploads photographs from the iPhone to Flickr. A separate service hosted by the application developer could monitor the Flickr upload and send a push notification through Apple when new comments are added to the photo on Flickr.

This push model is still somewhat controversial, as it only partially solves some of the needs of background applications. Still, knowing about the service as it comes to fruition will be important, especially in the context of the cooperative single-tasking ideals. This is because it frames each application as not only part of the user experience of the device, but as part of a network of interconnected services: mail, web, messaging, and any number of proprietary and tunneled protocols and services.

CHAPTER 6

Touch Patterns

The most famous feature of the iPhone and iPod Touch is the Multi-Touch interface. Multi-Touch allows a user to interact with a device using one or more fingers on a smooth, consistent physical screen. Touch-based interfaces have existed in prototypes and specialty devices for a while, but the iPhone and iPod Touch introduced the concept to the general consumer market. It's safe to say that the interaction pattern has proven very effective and popular, inspiring other companies to implement similar systems on their devices.

Any new interface requires updated patterns for accepting and handling input and for providing feedback to users. Apple has identified several simple and intuitive patterns not entirely dissimilar from those for traditional mouse use, but specialized for a Multi-Touch interface. Paired with the conceptual patterns and physical hardware are several libraries developers can use to manage user interaction. The currency of Multi-Touch programming is the `UITouch` class, which is one of many related classes in `UIKit`.

In Cocoa Touch applications, user input actions like button presses trigger *events*. The iPhone OS processes a related series of touches by grouping them into Multi-Touch sequences. Possible key events in a hypothetical sequence are listed here:

- One finger touches the device
- A second finger optionally touches the device
- One or both fingers move across the screen
- One or both fingers lift off the device
- A series of quick taps, such as a double-tap

The number of touch combinations that can make up a sequence seems endless. For this reason, it's important to examine established patterns and user expectations when deciding how to implement event management inside an application. In addition to sequences, touch accuracy and the visibility of "hot" controls or areas are vital to providing a good user experience. An application with buttons that are too small or too close together is likely to lead to frustration. This is also true of controls in areas that fingers or thumbs tend to block.

Touches and the Responder Chain

The class that represents touch events is the UITouch class. As a user interacts with the Multi-Touch interface, the operating system continually sends a stream of events to the dominant application. Each event includes information about all distinct touches in the current sequence. Each snapshot of a touch is represented by an instance of UITouch. The UITouch instance representing a given finger is updated through the sequence until it ends by all fingers being removed from the interface or by an external interruption.

UITouch Overview

As a user moves his finger across the screen, the current UITouch instances are updated to reflect several local (read-only) properties. The UITouch class is described in Figure 6-1.

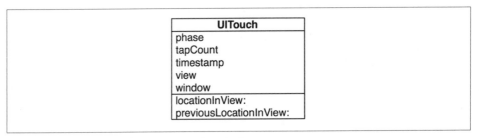

Figure 6-1. Public UITouch properties and methods

The following is a list of public properties of `UITouch:

tapCount
> The tapCount represents the number of quick, repeated taps associated with the UITouch instance.

timestamp
> The timestamp is the time when the touch was either created (when a finger touched the screen) or updated (when successive taps or fingertip movement occurred).

phase
> The phase value is a constant indicating where the touch is in its lifecycle. The phases correspond to: touch began, touch moved, touch remained stationary, touch ended, and touch canceled.

view
> The view property references the UIView in which the touch originated.

window
> Like the view property, the window property references the UIWindow instance in which the touch originated.

In addition to these properties, the UITouch class provides helpful methods for accessing the two-dimensional point (x, y) relative to a given UIView, representing both the current location and the location immediately preceding the current location. The locationIn View: and previousLocationInView: methods accept a UIView instance and return the point (as a CGPoint) in the coordinate space of that view.

UITouch instances are updated constantly, and the values change over time. You can maintain state by copying these properties into an appropriate structure of your choosing as the values change. You cannot simply copy the UITouch instance because UITouch doesn't conform to the NSCopying protocol.

The Responder Chain

Cocoa and Cocoa Touch applications both handle UI events by way of a *responder chain*. The responder chain is a group of *responder objects* organized hierarchically. A responder object is any object that inherits from UIResponder. Core classes in UIKit that act as responder objects are UIView, UIWindow, and UIApplication, in addition to all UIControl subclasses. Figure 6-2 illustrates the responder chain.

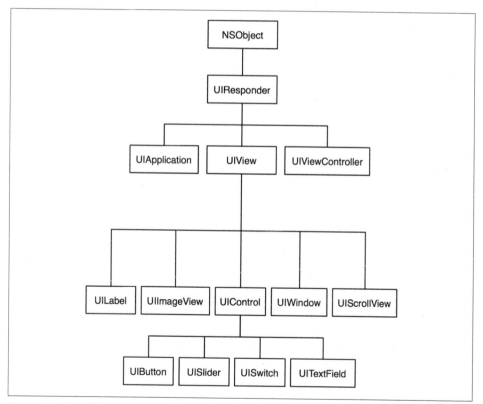

Figure 6-2. The UIKit responder chain

When a user interacts with the device, an *event* is generated by the operating system in response to the user interaction. An event in this case is an instance of the UIEvent class. Figure 6-3 shows the UIEvent class model.

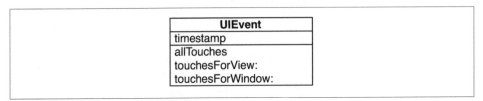

Figure 6-3. Public UIEvent properties and methods

Each new event moves up the responder chain from the most to least specific object. Not all descendants of UIResponder are required to handle all events. In fact, a responder object can ignore all events. If a responder object lacks a known event handler method for a specific event, the event will be passed up the chain until it is encountered by a responder object willing to handle it. That responder object can choose to pass the event on to the next responder in the chain for further processing, whether or not the responder object has acted on the event.

Becoming a responder object requires two steps:

1. Inherit from UIResponder or a descendant of UIResponder, such as UIView, UIButton, or UIWindow.

2. Override one of four touch-related event handler methods inherited from UIResponder.

The following list contains descriptions of UIResponder event handler methods:

- The touchesBegan:withEvent: method is called when one or more fingers touch the Multi-Touch surface. Very often, this will represent the initial contact in a sequence of single-finger touches. When objects enable support for multiple touches per sequence (such as with the familiar pinch-to-zoom gesture), this method may be called twice per sequence. To enable multiple touches per sequence, a responder object must declare that it wishes to receive multiple touches per sequence. This is done by sending a message, setMultipleTouchEnabled:, to the instance with an affirmative YES parameter.

    A frequent determination for the touchesBegan:withEvent: method is whether a touch is initial or supplemental in the sequence. The logic you implement for handling touches and gestures will often depend on state data around the entire sequence; therefore, you will want to initiate your data with an initial touch and only add to it for supplemental touches.

- The touchesMoved:withEvent: method is called when a finger moves from one point on the screen to another without being lifted. The event will be fired with each pass of the event loop, and not necessarily with each pixel-by-pixel movement. Though the stream is nearly constant, it's worth keeping in mind that the interval between calls is dependent upon the event loop and is thus technically variable.

This method is an excellent point at which to record the location of the full set of UITouch instances delivered with the UIEvent parameter. The touches Moved:withEvent: method is called very frequently during a touch sequence—often hundreds of times per second—so be careful of using it for expensive work.

- The touchesEnded:withEvent: method is invoked when both fingers (or one, in a single-touch application) are lifted from the Multi-Touch screen. If your responder object accepts multiple touches, it may receive more than one touchesEnded:with Event: message during the touch sequence, as a second finger makes contact and then leaves the screen.

As with the touchesCancelled:withEvent: method, you will often perform the bulk of your logic and cleanup operations when this message is received.

- The touchesCancelled:withEvent: method is called when the touch sequence is canceled by an external factor. Interruptions from the operating system, such as a warning for low memory or an incoming phone call, are fairly common. As you'll see in this chapter, the art of managing touches often includes managing state around touch sequences, and persisting and updating that state across events. It's therefore important to use both the touchesEnded:withEvent: and the touches Cancelled:withEvent: methods to perform any operations that manage state. For example, deleting a stack of UITouch objects and committing/undoing a graphics transaction are possible cleanup operations.

Each event contains the full set of UITouch instances included in the Multi-Touch sequence of which it is a part. Each UITouch contains a pointer to the UIView in which the touch event was generated. Figure 6-4 illustrates the relationship.

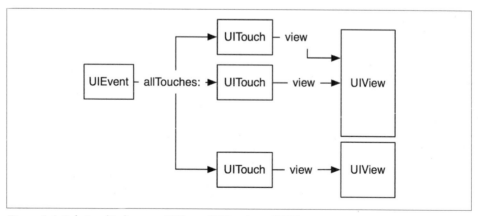

Figure 6-4. Relationship between UIEvent, UITouch, and UIView

Touch Accuracy

An instance of UITouch exposes its location as a two-dimensional CGPoint value. Each CGPoint represents an (x, y) pair of float values. Clearly, even the tiniest fingertip is much larger than a single point on the screen. The iPhone does a great job of training users to expect and accept the approximate fidelity that results from translating a physical touch to a single point in the coordinate space of a view. Still, developers with an appreciation for user experience should pay attention to the perception of accuracy. If a user feels that input results in a loss of precision, frustration is a likely outcome.

The main considerations for touch accuracy are:

- The size of touchable objects
- The shape of touchable objects
- The placement of touchable objects in relation to one another
- The overlapping of touchable objects

Size

The size of touchable objects is an interesting problem. One of the more curious facets of a portable touch interface is that the main input device (a finger) also obfuscates the feedback mechanism (the screen). Touching a control, such as a button, should provide users with visual feedback to provide a confirmation that their intentions have been communicated to the device. So how does Apple address this issue in UIKit? They attack the issue from many angles.

First, many controls are quite large. By displaying buttons that span approximately 80% of the width of the screen, Apple guarantees that users can see portions of the button in both its highlighted and its touched state. The passive confirmation mechanism works very well. Figure 6-5 shows the device emulator included in the iPhone SDK with the Contacts application running. The "Delete Contact" and "Cancel" buttons are good examples of very prominent, large controls.

In addition to expanding the visible area of controls into the periphery, Apple has bolstered the ambient feedback mechanism by changing the hit area of these controls for drags. In desktop Cocoa applications, interaction is canceled when the mouse is dragged outside the visible boundaries of the view handling the event. With Cocoa Touch controls on the iPhone OS, Apple drastically expands the "hot" area of the control on touch. This means that touches require a certain level of accuracy, but the chance of accidentally dragging outside of a control and inadvertently canceling a touch sequence is lessened. This allows users to slightly drag away from a control to visually confirm their input. This implementation pattern is free with standard controls and in many cases with subclasses. When drawing your own views and managing your own hit test logic, you should attempt to copy this functionality to ensure compliance with the new muscle memory users acquire on the iPhone OS. Figure 6-6 displays three similar controls. The first is standard; the second displays the hot area for receiving touches; and the third displays the virtual hit area for active touches. Dragging outside of the area highlighted in the figure cancels the selection.

Figure 6-5. Examples of large buttons in the Contacts application

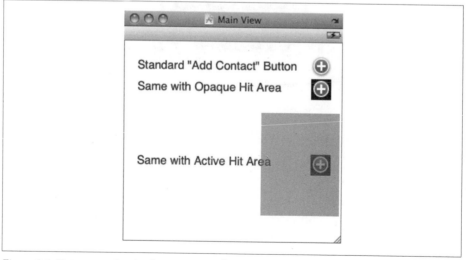

Figure 6-6. Hot area and active hot area examples

The onscreen keyboard has an elegant solution to the problem of touch area. The size of each button in the keyboard is smaller than the typical adult fingertip. Since the keyboard layout is a standard QWERTY configuration, users are familiar with the location of each key. But because the keyboard is displayed on screen, the standard "home row" finger positions and ingrained muscle memory can't help accuracy. Apple allows users to confirm the input of each key by briefly expanding the key graphics above the touch location. This pattern is also used in an enhanced form for special keys, such as the *.com* key added conditionally to the keyboard when the first responder field represents a URL. Figure 6-7 illustrates the touch-and-hold control style.

Figure 6-7. A standard touch-and-hold control

You can use virtual hit areas to enlarge the hot area for a control without changing the visual interface. You can override the `pointInside:withEvent:` or `hitTest:withEvent:` method to create a virtual hit area. This method is called for a `UIView` by its `superview` property as a part of the responder chain. Returning a `NO` value from these methods causes the responder chain to move on to the next responder object in the chain. Returning `YES` allows the responder object to handle the event and terminate the trip up the responder chain. Creating a virtual hit area may be as simple as returning `YES` for points outside the visible boundaries of the view.

The following example creates an enlarged virtual hit area:

```objc
// HotView.h
#import <UIKit/UIKit.h>

@interface HotView : UIView {
    BOOL hot;
}

@end

// HotView.m
#import "HotView.h"

@implementation HotView

- (id)initWithFrame:(CGRect)frame
{
    if (self = [super initWithFrame:frame]) {
        hot = true;
    }
    return self;
}

#define MARGIN_SIZE 10.0
#define DRAGGING_MARGIN_SIZE 40.0

- (BOOL) point:(CGPoint)point insideWithMargin:(float)margin
{
    CGRect rect = CGRectInset(self.bounds, -margin, -margin);
    return CGRectContainsPoint(rect, point);
}

- (BOOL) pointInside:(CGPoint)point withEvent:(UIEvent *)event
{
    float phasedMargin;
    UITouch *touch = [[event touchesForView:self] anyObject];
    if(touch.phase != UITouchPhaseBegan){
        phasedMargin = DRAGGING_MARGIN_SIZE;
    }else{
        phasedMargin = MARGIN_SIZE;
    }

    if([self point:point insideWithMargin:phasedMargin]){
        return YES;
    }else{
        return NO;
    }
}

- (void) touchesBegan:(NSSet *)touches withEvent:(UIEvent *)event
{
    NSLog(@"Touches began.");
    hot = YES;
}
```

```
- (void) touchesMoved:(NSSet *)touches withEvent:(UIEvent *)event
{
    if(hot == NO) return;
    CGPoint point = [[touches anyObject] locationInView:self];
    if([self point:point insideWithMargin:DRAGGING_MARGIN_SIZE] == false){
        [self.nextResponder touchesBegan:touches withEvent:event];
        hot = NO;
    }

    NSLog(@"Touch moved.");
}

- (void)touchesEnded:(NSSet *)touches withEvent:(UIEvent *)event
{
    if(hot == NO) return;
    NSLog(@"Touches ended.");
    hot = YES;
}

- (void)touchesCancelled:(NSSet *)touches withEvent:(UIEvent *)event
{
    if(hot == NO) return;
    NSLog(@"Touches cancelled.");
    hot = YES;
}

@end
```

Shape

Designing touch-enabled views with irregular shapes is appropriate in many applications. Luckily, Cocoa Touch application developers can use any of several strategies for deciding when a custom view should handle a touch sequence.

When a touch is being handled by the view hierarchy, the hitTest:withEvent: message is sent to the topmost UIView in the view hierarchy that can handle the touch event. The top view then sends the pointInside:withEvent: message to each of its subviews to help divine which descendant view should handle the event.

You can override pointInside:withEvent: to perform any logic required by your custom UIView subclass.

For example, if your view renders itself as a circle centered inside its bounds and you'd like to ignore touches outside the visible circle, you can override pointInside:withEvent: to check the location against the radius of the circle:

```
- (BOOL)pointInside:(CGPoint)point withEvent:(UIEvent *)event
    // Assume the view/circle is 100px square
    CGFloat x = (point.x - 50.0) / 50.0;
    CGFloat y = (point.y - 50.0) / 50.0;
    float h = hypot(x, y);
    return (h < 1.0);
}
```

If you have an irregular shape that you've drawn with `CoreGraphics`, you can test the `CGPoint` against the bounds of that shape using similar methods.

In some cases, you may have an image in a touch-enabled `UIImageView` with an alpha channel and an irregular shape. In such cases, the simplest means of testing against the shape is to compare the pixel at the `CGPoint` against a bitmap representation of the `UIImageView`. If the pixel in the image is transparent, you should return `NO`. For all other values, you should return `YES`.

Placement

The placement of views in relation to one another affects usability and perception of accuracy as much as the size of controls. The iPhone is a portable Multi-Touch device and thus lends itself to accidental or imprecise user input. Applications that assist users by attempting to divine their intentions probably gain an advantage over competing applications with cluttered interfaces that demand focus and precision from users. Virtual hit areas for untouched states are difficult or impossible to use when views are very close together.

When two views touch one another and a finger touches the edges of both, the view most covered by the fingertip will act as the first responder in the responder chain and receive the touch events. Regardless of the view in which the touch originated, you can get the location of a `UITouch` instance in the coordinate system of any `UIView`, or in the `UIWindow`. You can program your views in a way that maintains encapsulation when a `UITouch` instance is processed:

```
// Get the location of a UITouch (touch) in a UIView (viewA)
CGPoint locationInViewA = [touch locationInView:viewA];

// Get the location of a UITouch (touch) in a UIView (viewB)
CGPoint locationInViewB = [touch locationInView:viewB];

// Get the location of a UITouch (touch) in the UIView that
// is the current responder
CGPoint locationInSelf = [touch locationInView:self];

// Get the location of a UITouch (touch) in the main window
CGPoint locationInWindow = [touch locationInView:nil];
```

Depending on the shape and meaning of the view handling a touch event, you should consider placement in relation to a fingertip when appropriate. A great example of this is when dragging a view under a fingertip. If you require precision when users drag a view around the screen, you can improve the user experience by positioning the element slightly above the touch instead of centering it under the touch:

```
- (void)touchesMoved:(NSSet *)touches withEvent:(UIEvent *)event
{
    UITouch *touch = [touches anyObject];
    CGPoint location = [touch locationInView:self];
```

```
    // Positioning directly under the touch
    self.center = location;
float halfHeight = self.frame.size.height * 0.5;
    CGpoint betterLocation = CGPointMake(location.x, (location.y - halfHeight));
    // Positioning slightly above the touch
    self.center = betterLocation;
}
```

Overlapping Views

Designing a user experience that allows elements to overlap each other on the z-axis[*]
presents a few key challenges:

- If the overlapping elements are movable by users or animations, care should be
  taken to prevent any single element from fully covering another element. If such
  behavior is expected, users should be given some means of easily accessing underlying elements.

- If an overlapping area has an irregular shape, the desired behavior is probably to
  restrict the hit area to the shape and not to the bounding rectangle. Doing so allows
  touch events to pass "through" the bounding rectangle of the top element to the
  bottom element.

- Enlarged virtual hit areas are more difficult to program when touchable views
  overlap because the logic for passing touch events down the stack could conflict
  with the logic that facilitates virtual hit areas.

Apple recommends not allowing sibling views to overlap one another for both usability
and performance reasons. You can find additional information on overlapping UIKit
views in the iPhone Application Programming Guide, which can be found online at
*http://developer.apple.com/iphone/library/documentation/iPhone/Conceptual/iPhoneOS
ProgrammingGuide/Introduction/Introduction.html*.

Detecting Taps

So far, this chapter has focused on the conceptual side of Multi-Touch programming.
The remainder of the chapter will focus on example code showing how to detect and
use the main types of touch sequence.

Detecting Single Taps

Single taps are used by standard buttons, links (in browsers and the SMS application),
and many other UIControl subclasses. They are also used by the iPhone OS to launch
applications. Users touch elements on the screen to communicate intent and, in doing

---

[*] 3D has three axes: x, y, and z. When applied to 2D displays, the z-axis is—to your eyes—the surface of the
screen. So when things overlap, it occurs on the z-axis.

so, expect a response. On the Home screen, the response is to launch an application. With buttons, a specific action is usually expected: search, close, cancel, clear, accept.

Single taps are trivial to detect. The simplest method is to assign an action to a `UIControl` subclass (versus a custom `UIView` subclass). This sends a specific message to a given object. For a given `UIControl`, send the `addTarget:action:forControlEvents:` message with appropriate parameters to assign a receiving target and action message for any number of control events. This example assumes a `UIButton` instance in a `UIView` subclass with the instance variable name `button`:

```
- (void) awakeFromNib
{
    [super awakeFromNib];
    [button addTarget:self action:@selector(handleButtonPress:)
        forControlEvents:UIControlEventTouchDown];
}

- (IBAction) handleButtonPress:(id)sender
{
    NSLog(@"Button pressed!");
}
```

For responder objects that are not descendants of `UIControl`, you can detect single taps within the `touchesBegan:withEvent:` handler:

```
- (void) touchesBegan:(NSSet *)touches withEvent:(UIEvent*)event
{
    UITouch *touch = [touches anyObject];
    NSUInteger numTaps = [touch tapCount];
    NSLog(@"The number of taps was: %i", numTaps);
    if(numTaps == 1){
        NSLog(@"Single tap detected.");
    }else{
        //    Pass the event to the next responder in the chain.
        [self.nextResponder touchesBegan:touches withEvent:event];
    }
}
```

Detecting Multiple Taps

You can handle multiple taps similarly to single taps. The `UITouch tapCount` property will increment appropriately to reflect the number of taps within the same sequence. Most computer interaction systems use single and double tap patterns. For special cases, such as certain games, you may wish to allow users to use triple taps—or endless taps. If a sufficient pause between taps occurs, the operating system treats new taps as part of a new sequence. If you'd like to handle repeated tapping with longer pauses, you should write logic that maintains state between multiple touch sequences and treats them as members of the same series within the temporal boundaries you set:

```
- (void) touchesBegan:(NSSet *)touches withEvent:(UIEvent*)event
{
    UITouch *touch = [touches anyObject];
```

```
    NSUInteger numTaps = [touch tapCount];
    NSLog(@"The number of taps was: %i", numTaps);
    if(numTaps > 1){
        NSLog(@"Multiple taps detected.");
    }
}
```

Detecting Multiple Touches

Handling multiple touches in a sequence is different from handling multiple taps for a single touch. Each `UIEvent` dispatched up the responder chain can contain multiple `UITouch` events—one for each finger on the screen. You can derive the number of touches by counting the `touches` argument to any of the touch event handlers:

```
- (void) touchesBegan:(NSSet *)touches withEvent:(UIEvent*)event
{
    int numberOfTouches = [touches count];
    NSLog(@"The number of fingers on screen: %i", numberOfTouches);
}
```

Handling Touch and Hold

An interesting control present in the onscreen keyboard is the *.com* button that appears when a URL entry field has focus. Quickly tapping the button like any other key inserts the string ".com" into the field. Tapping on the control and holding it down for a moment causes a new subview to appear with a set of similar buttons representing common top-level domain name parts, such as *.net* and *.org*.

To program a similar touch-and-hold control, you need to detect that a touch has begun and that an appropriate amount of time has passed without the touch being completed or canceled. There are many ways to do so, but the use of a timer is a simple solution:

```
// Expander.h
@interface Expander : UIView {
    UIView *expandedView;
    NSTimer *timer;
}

@end

// Expander.m
import "Expander.h"

@interface Expander ()
- (void)stopTimer;
- (void)close;
- (void)expand:(NSTimer *)theTimer;
@end

@implementation Expander
```

```objc
- (id)initWithFrame:(CGRect)frame
{
    if(self = [super initWithFrame:frame]){
        self.frame = CGRectMake(0.0, 0.0, 40.0, 40.0);
        self.backgroundColor = [UIColor redColor];

        expandedView = [[UIView alloc] initWithFrame:CGRectZero];
        expandedView.backgroundColor = [UIColor greenColor];
        expandedView.frame = CGRectMake(-100.0, -40.0, 140.0, 40.0);
        expandedView.hidden = YES;
        [self addSubview:expandedView];
    }
    return self;
}

- (void)touchesBegan:(NSSet *)touches withEvent:(UIEvent *)event
{
    [self stopTimer];
    timer = [NSTimer scheduledTimerWithTimeInterval:1.0
                                    target:self
                                    selector:@selector(expand:)
                                    userInfo:nil
                                    repeats:NO];
    [timer retain];
}

- (void)touchesEnded:(NSSet *)touches withEvent:(UIEvent *)event
{
    [self stopTimer];
    [self close];
}

- (void)touchesCancelled:(NSSet *)touches withEvent:(UIEvent *)event
{
    [self stopTimer];
    [self close];
}

- (void)stopTimer
{
    if([timer isValid]){
        [timer invalidate];
    }
}

- (void)expand:(NSTimer *)theTimer
{
    [self stopTimer];
    expandedView.hidden = NO;
}

- (void)close
{
    expandedView.hidden = YES;
}
```

```
- (void)dealloc
{
    [expandedView release];
    [super dealloc];
}

@end
```

Handling Swipes and Drags

A UITouch instance persists during an entire drag sequence and is sent to all event handlers set up in a UIView. Each instance has mutable and immutable properties that are relevant to gesture detection.

As a finger moves across the screen, its associated UITouch is updated to reflect the location. The coordinates of the location are stored as a CGPoint and are accessible by way of the locationInView: method of the UIView class.

Dragging a view is simple. The following example shows the implementation of a simple UIView subclass, Draggable. When handling a touchesMoved:withEvent: message, a Draggable instance will position itself at the point of a touch relative to the coordinate space of its superview:

```
@implementation Draggable

- (id)initWithFrame:(CGRect)frame
{
    if (self = [super initWithFrame:frame]) {
        self.backgroundColor = [UIColor redColor];
    }
    return self;
}

- (void) touchesBegan:(NSSet *)touches withEvent:(UIEvent *)event
{
    NSLog(@"Touched.");
}

- (void) touchesMoved:(NSSet *)touches withEvent:(UIEvent *)event
{
    NSLog(@"Dragged.");
    UITouch *touch = [touches anyObject];
    CGPoint location = [touch locationInView:self.superview];
    self.center = location;
}

@end
```

Swipe detection is slightly more complex than drag management. In the iPhone Application Programming Guide, Apple recommends a strategy for detecting swipes that leads to consistent user behavior across applications. Conforming to the standard set by Apple helps improve user experience because it helps build and takes advantage of muscle memory. For example, UIKit includes built-in support for detecting swipes across table cells, prompting users with a button to delete. Mapping the swipe-to-delete gesture in default applications—and in UIKit as a free feature—helps to "train" users that the swipe is a dismissive gesture. This carries over to other uses of the swipe gesture. Another example is the Photos application. Users can swipe across a photo when viewing a gallery. The gesture will dismiss the current photo and, depending on the swipe direction, transition the next or previous photo into place.

You can leverage the swipe to perform your own equivalent of dismissal:

```
// MainView.h
@interface MainView : UIView {
    Spinner *spinner;
}

// MainView.m
@interface MainView (PrivateMethods)

- (void)transformSpinnerWithFirstTouch:(UITouch *)firstTouch
    andSecondTouch:(UITouch *)secondTouch;
- (CGFloat)distanceFromPoint:(CGPoint)fromPoint toPoint:(CGPoint)toPoint;
- (CGPoint)vectorFromPoint:(CGPoint)firstPoint toPoint:(CGPoint)secondPoint;

@end

@implementation MainView

- (void)awakeFromNib
{
    self.multipleTouchEnabled = YES;
    spinner = [[Spinner alloc] initWithFrame:CGRectMake(0.0, 0.0, 50.0, 50.0)];
    spinner.center = self.center;
    [self addSubview:spinner];
}

- (void)touchesMoved:(NSSet *)touches withEvent:(UIEvent *)event
{
    if([touches count] != 2){
        return;
    }
    NSArray *allTouches = [touches allObjects];
    UITouch *firstTouch = [allTouches objectAtIndex:0];
    UITouch *secondTouch = [allTouches objectAtIndex:1];
    [self transformSpinnerWithFirstTouch:firstTouch andSecondTouch:secondTouch];
}

- (void)transformSpinnerWithFirstTouch:(UITouch *)firstTouch
    andSecondTouch:(UITouch *)secondTouch
{
    CGPoint firstTouchLocation = [firstTouch locationInView:self];
    CGPoint firstTouchPreviousLocaion = [firstTouch previousLocationInView:self];
    CGPoint secondTouchLocation = [secondTouch locationInView:self];
    CGPoint secondTouchPreviousLocation = [secondTouch previousLocationInView:self];
```

```
    CGPoint previousDifference = [self vectorFromPoint:firstTouchPreviousLocaion
        toPoint:secondTouchPreviousLocation];
    CGAffineTransform newTransform =
        CGAffineTransformScale(spinner.transform, 1.0, 1.0);
    CGFloat previousRotation = atan2(previousDifference.y, previousDifference.x);
    CGPoint currentDifference = [self vectorFromPoint:firstTouchLocation
        toPoint:secondTouchLocation];
    CGFloat currentRotation = atan2(currentDifference.y, currentDifference.x);
    CGFloat newAngle = currentRotation - previousRotation;
    newTransform = CGAffineTransformRotate(newTransform, newAngle);
    spinner.transform = newTransform;
}

- (CGFloat)distanceFromPoint:(CGPoint)fromPoint toPoint:(CGPoint)toPoint
{
    float x = toPoint.x - fromPoint.x;
    float y = toPoint.y - fromPoint.y;
    return hypot(x, y);
}

- (CGPoint)vectorFromPoint:(CGPoint)firstPoint toPoint:(CGPoint)secondPoint
{
    CGFloat x = firstPoint.x - secondPoint.x;
    CGFloat y = firstPoint.y - secondPoint.y;
    CGPoint result = CGPointMake(x, y);
    return result;
}


- (void) dealloc
{
    [spinner release];
    [super dealloc];
}

@end
```

Handling Arbitrary Shapes

The Multi-Touch interface allows developers to create interaction patterns based on simple taps, drags, and flicks. It also opens the door for more complex and engaging interfaces. We've seen ways to implement taps (single and multiple) and have explored dragging view objects around the screen. Those examples conceptually bind a fingertip to object in space, creating an interface through the sense of touch, or haptic experience. There is another way of thinking of touches in relation to user interface objects that is a little more abstract, but nonetheless compelling to users.

The following example creates an interface that displays a grid of simple tiles, as shown in Figure 6-8. Each tile has two states: on and off. When a user taps a tile, it toggles the state and updates the view to use an image that correlates to that state. In addition to tapping, a user can drag over any number of tiles, toggling them as the touch moves in and out of the bounds of the tile.

Figure 6-8. Sample tile-based application

Clicking the "Remove" button at the bottom of the screen removes all tiles in the selected state and triggers a short animation that repositions the remaining tiles:

```
// Board.h
#import <UIKit/UIKit.h>
#import "Tile.h"

@interface Board : UIView {
    NSMutableArray *tiles;
    Tile *currentTile;
    BOOL hasTiles;
}

@property (nonatomic, retain) NSMutableArray *tiles;
@property (nonatomic, assign) BOOL hasTiles;

- (void)clear;
- (IBAction)removeSelectedTiles;
- (void)addTile;
- (void)removeTile:(Tile *)tile;

@end
```

```objc
// Board.m
#import "Board.h"

@interface Board (PrivateMethods)

- (void)setup;
- (void)toggleRelevantTilesForTouches:(NSSet *)touches
    andEvent:(UIEvent *)event;

@end

@implementation Board

@synthesize tiles, hasTiles;

- (id)initWithFrame:(CGRect)frame
{
    if(self = [super initWithFrame:frame]){
        [self setup];
    }
    return self;
}

- (void)addTile
{
    [tiles addObject:[[[Tile alloc] init] autorelease]];
}

- (void)removeTile:(Tile *)tile
{
    if([tiles containsObject:tile]){
        [tiles removeObject:tile];
        [tile disappear];
    }
    if([tiles count] < 1){
        self.hasTiles = NO;
    }else{
        self.hasTiles = YES;
    }
}

- (void)clear
{
    Tile *tile;
    for(tile in tiles){
        [self removeTile:tile];
    }
    self.hasTiles = NO;
}

- (void)willRemoveSubview:(UIView *)subview
{
    [self removeTile:(Tile *)subview];
}
```

```
- (IBAction)removeSelectedTiles
{
    Tile *tile;
    NSArray *tilesSnapshot = [NSArray arrayWithArray:tiles];
    for(tile in tilesSnapshot){
        if(tile.selected){
            [self removeTile:tile];
        }
    }
    if([tiles count] < 1){
        self.hasTiles = NO;
    }else{
        self.hasTiles = YES;
    }
}


#define NUM_COLS        4
#define NUM_ROWS        5
#define MARGIN_SIZE        2
#define TILE_COUNT        NUM_COLS * NUM_ROWS

- (void)setup
{
    if(tiles == nil){
        tiles = [NSMutableArray arrayWithCapacity:TILE_COUNT];
        [tiles retain];
    }
    for(int i = 0; i < TILE_COUNT; i++){
        [self addTile];
    }
    self.backgroundColor = [UIColor whiteColor];
    [self setNeedsDisplay];
}

- (void)layoutSubviews
{
    Tile *tile;
    int currentRow = 0;
    int currentColumn = 0;
    int i = 0;
    float tileSize = (320.0/NUM_COLS) - (MARGIN_SIZE * 1.25);
    float x, y;
    for(tile in tiles){
        //    Lay out the tile at the given location
        [self addSubview:tile];
        x = (currentColumn * tileSize) + (MARGIN_SIZE * (currentColumn + 1));
        y = (currentRow * tileSize) + (MARGIN_SIZE * (currentRow + 1));
        [tile appearWithSize:CGSizeMake(tileSize, tileSize)
            AtPoint:CGPointMake(x, y)];
        if(++i % 4 == 0){
            currentRow++;
            currentColumn = 0;
        }else{
            currentColumn++;
```

```
        }
        [tile setNeedsDisplay];
    }
}

- (void)touchesBegan:(NSSet *)touches
    withEvent:(UIEvent *)event
{
    currentTile = nil;
    [self toggleRelevantTilesForTouches:touches andEvent:event];
}

- (void)touchesEnded:(NSSet *)touches
    withEvent:(UIEvent *)event
{
    currentTile = nil;
}

- (void)touchesCancelled:(NSSet *)touches
    withEvent:(UIEvent *)event
{
    currentTile = nil;
}

- (void)touchesMoved:(NSSet *)touches
    withEvent:(UIEvent *)event
{
    [self toggleRelevantTilesForTouches:touches andEvent:event];
}

- (void)toggleRelevantTilesForTouches:(NSSet *)touches
    andEvent:(UIEvent *)event
{
    UITouch *touch = [touches anyObject];
    Tile *tile;
    CGPoint location;
    for(tile in tiles){
        location = [touch locationInView:tile];
        if([tile pointInside:location withEvent:event]){
            //    if the touch is still over the same tile, get out
            if(tile == currentTile){
                continue;
            }
            [tile toggleSelected];
            currentTile = tile;
        }
    }
}

- (void)dealloc {
    [tiles release];
    [currentTile release];
    [super dealloc];
}
```

```
@end

// Tile.h
#import <UIKit/UIKit.h>

@interface Tile : UIView {
    BOOL selected;
    BOOL hasAppeared;
    UIImageView *backgroundView;
}

@property (nonatomic, assign) BOOL selected;

- (void)toggleSelected;
- (void)disappear;
- (void)appearWithSize:(CGSize)size
    AtPoint:(CGPoint)point;

@end

// Tile.m
#import "Tile.h"

@implementation Tile

@synthesize selected;

- (id)init
{
    if (self = [super init]) {
        self.backgroundColor = [UIColor clearColor];
        backgroundView = [[UIImageView alloc]
            initWithImage:[UIImage imageNamed:@"on.png"]];
        [self addSubview:backgroundView];
        [self sendSubviewToBack:backgroundView];
        self.selected = NO;
        hasAppeared = NO;
    }
    return self;
}

- (void)moveToPoint:(CGPoint)point
{
    [UIView beginAnimations:nil context:nil];
    [UIView setAnimationDuration:0.5];
    CGRect frame = self.frame;
    frame.origin = point;
    self.frame = frame;
    [UIView commitAnimations];
}

- (void)appearWithSize:(CGSize)size AtPoint:(CGPoint)point
{
    //    If it's new, have it 'grow' into being
    if(!hasAppeared){
```

```
        CGRect frame = self.frame;
        frame.origin = point;
        frame.size = size;
        self.frame = frame;

        //    Shrink it
        CGAffineTransform shrinker =
            CGAffineTransformMakeScale(0.01, 0.01);
        self.transform = shrinker;

        //    Start the animations transaction
        [UIView beginAnimations:nil context:nil];
        [UIView setAnimationDuration:0.5];

        //    Grow it
        CGAffineTransform grower =
            CGAffineTransformScale(self.transform, 100.0, 100.0);
        self.transform = grower;

        //    Commit the transaction
        [UIView commitAnimations];

        //    Flag that I have been on screen
        hasAppeared = YES;
    }else{
        [self moveToPoint:point];
    }
}

- (void)touchesBegan:(NSSet *)touches
    withEvent:(UIEvent *)event
{
    UITouch *touch = [touches anyObject];
    if([touch tapCount] == 2){
        [self removeFromSuperview];
    }else{
        [self.nextResponder touchesBegan:touches withEvent:event];
        return;
    }

}

- (void)disappear
{
    [UIView beginAnimations:nil context:nil];
    [UIView setAnimationDuration:0.5];
    CGAffineTransform transform =
        CGAffineTransformMakeScale(.001, .001);
    self.transform = transform;
    [UIView commitAnimations];
}

- (void)toggleSelected
{
    self.selected = !self.selected;
```

```
    if(self.selected){
        backgroundView.image = [UIImage imageNamed:@"off.png"];
    }else{
        backgroundView.image = [UIImage imageNamed:@"on.png"];
    }
}

- (void)drawRect:(CGRect)rect
{
    self.bounds = self.frame;
    backgroundView.frame = self.bounds;
}

- (void)dealloc
{
    [backgroundView release];
    [super dealloc];
}


@end
```

CHAPTER 7

Interaction Patterns and Controls

User interface design is an important part of application development because the UI is the face of an application. In fact, to most users, the interface *is* the application. Designers and developers spend a lot of time deciding how common interface elements should look and how interaction should occur. Designers of Cocoa or Cocoa Touch applications can take advantage of the Cocoa UI frameworks to rapidly prototype and adapt their interfaces. In addition to the frameworks, Apple provides developers with descriptions of interaction patterns for common problems in the Human Interface Guidelines.

Application Interaction Patterns

The View Controller Programming Guide for iPhone OS, included in the iPhone SDK documentation, explores several high-level design patterns for user interaction. Apple created semantics for describing the overall style of interaction for a given screen or set of screens in an application. Most applications will implement designs that can be described according to this vocabulary. Designers with an understanding of common patterns can use that knowledge to plan interfaces according to user needs.

Command Interfaces

A command interface presents users with a toolbar containing one or more buttons that represent executable actions. Command interfaces typically don't use view controllers, and instead wire actions for buttons to other objects directly. You add a UIToolbar to your application to implement a command interface.

Command interfaces might be appropriate if:

- You have a very small, finite set of actions to display.
- You have an editable view, such as a drawing surface, and want a set of tools or actions.

- Your application is more robust than a simple, read-only utility, but has one main screen on which users operate.

Figure 7-1 shows examples of command interfaces, including Face Melter, Notes, and Things.

Figure 7-1. Command interfaces in Face Melter, Notes, and Things

Radio Interfaces

Radio interfaces display a set of buttons that switch between views when tapped. The buttons display on a tab bar at the bottom of the window. Each tap swaps to a different view without animating between the views. This type of interface works well for displaying non-hierarchical data. You can use a `UITabBar` to create a radio interface.

Radio interfaces might be appropriate if:

- You have a set of related but disparate screens. If your screens aren't related in nature, you should consider building multiple applications to adhere to the concept of *cooperative single-tasking*, described in Chapter 5.
- Your views are siblings. That is, they don't represent different levels of a hierarchy of data, but rather various views into data that may or may not overlap.
- You have a small set of closely related subsets of functionality that can be accessed in any order. Essentially, each view requires a separate main view controller, so the partitioning functionality should consider the architecture.

Figure 7-2 shows examples of radio interfaces, including Clock and Foursquare.

Figure 7-2. Radio interfaces in Clock and Foursquare

Navigation Interfaces

Navigation interfaces display a hierarchy of objects. Users tap controls to move toward greater specificity. The `UINavigationController` class is typically used to navigate a hierarchy of `UIViewController` instances. Changing between views animates the more specific view in from the right, while the less specific view moves out toward the left. Moving back up the hierarchy animates the views in the other direction.

Navigation interfaces might be appropriate if:

- You have a set of hierarchical data or functionality that you'd like to allow users to traverse. If your data fits any tree-like data structure, a navigation interface is likely appropriate, and follows standards established by Apple as part of the Cocoa Touch platform.

Figure 7-3 shows an example of a navigation interface in OmniFocus and a custom navigation interface in Birdfeed.

Modal Interfaces

A modal interface presents a supplemental view on top of a view. This is most useful when presenting secondary screens, such as an editing screen. Modal interfaces are similar to navigation interfaces in that they animate transitions between views. Navigation interfaces transition through a stack of `UIViewControllers` by animating them horizontally. Modal interfaces differ by transitioning new views onto the top of the stack by animating them vertically.

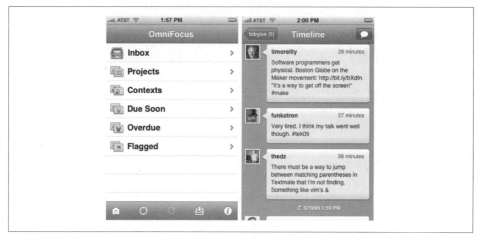

Figure 7-3. Navigation interfaces in OmniFocus and Birdfeed

Modal interfaces might be appropriate if:

- You need to provide a workflow for operating on data in multiple steps.
- You need to provide an optional editable interface for an onscreen view.
- You have a secondary set of functionality that users can optionally engage while remaining in the context of the current view. For example, you can use a modal interface if you want to trigger the camera to take a photograph or choose a contact from your address book, and then provide the resulting data to the current view.

Figure 7-4 shows two modal interfaces in Birdfeed, one for accessing third-party services for a Twitter account and another for sending a message to a user.

Figure 7-4. Modal interfaces in Birdfeed

Combination Interfaces

Radio, navigation, and modal interfaces are not mutually exclusive, and are in fact often coupled to provide an interface that presents multiple sets of hierarchical data for navigation. The use of combination interfaces is very common. Navigation and modal interfaces are easily combined to create an application that can navigate through one of several hierarchical datasets. You can recognize the modal-navigation interface by a tab bar on the bottom of the screen that switches between table views at the top of the screen.

Combination interfaces might be appropriate if:

- You have a complex set of data or functionality that would benefit from multiple patterns.
- Users interact with your application using different modes for the same dataset, such as charting, reading, and editing statistics.

Figure 7-5 shows an example of a combination interface using the TED application. The TED application allows users to browse content related to the annual TED conference.

Figure 7-5. A combination radio and navigation interface in TED

UIControl Classes

Cocoa Touch provides a robust set of user interface controls that can be used in iPhone and iPod Touch applications. The controls included in UIKit help ensure consistency across applications by establishing and conforming to a set of defined interaction patterns. Leveraging familiar interaction patterns in your applications can increase usability by reducing the burden of learning the intricacies of a new application. For lightweight, task-oriented mobile applications, a shallow learning curve allows users to find value quickly.

User interface elements can be split into two main categories: those that accept user input and those that do not. Elements that respond to user input are called controls. Examples of controls are buttons, sliders, and switches. Non-control elements include activity indicators and status bars.

Standard controls are descendants of UIControl and implement a special mechanism for sending messages in response to user interaction. As with many standardized aspects of Cocoa, this allows developers to rapidly evolve their application, changing the types of controls on the screen or altering the backend code that responds to user interaction.

The UIControl class is the foundation for all of the standard buttons, switches, and sliders. Figure 7-6 shows the UIControl class tree.

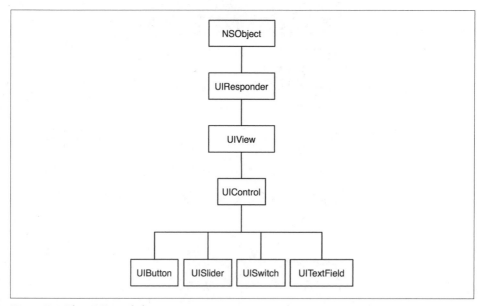

Figure 7-6. The UIControl class tree

The Target-Action Mechanism

In Chapter 6, we explored custom UIView programming and worked with UITouch for Multi-Touch interaction. Those patterns are perfect when the goal is to arbitrarily handle raw touch events. With control programming, the goal is to develop a (typically reusable) interface object that can report user intent in a clear, uniform way. Cocoa Touch provides a very simple mechanism for just this purpose.

The *target-action mechanism* allows instances of Objective-C classes to be registered with a UIControl descendant and messaged when certain control events occur, such as a touch or drag. Instances are registered as *targets* of the message, and they supply an *action* to handle the event.

Figure 7-7 illustrates a simple target-action sequence.

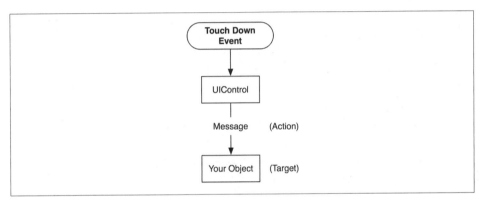

Figure 7-7. Conceptual messaging chain of the target-action mechanism

Types of Control Events

Objects register to receive notifications from controls using *control events*. The UIControl header defines keys that identify control events.

Table 7-1 lists the UIControlEvent values defined in UIControl.h.

Table 7-1. UIControlEvent listing

Event name	Event source
UIControlEventTouchDown	Fingertip touching the control
UIControlEventTouchDownRepeat	Fingertip touching multiple times, such as a double-tap
UIControlEventTouchDragInside	Fingertip dragging within the bounds of the control
UIControlEventTouchDragOutside	Fingertip dragging outside the bounds of the control
UIControlEventTouchDragEnter	Fingertip dragging from outside the bounds to inside the bounds of the control

Event name	Event source
UIControlEventTouchDragExit	Fingertip dragging from inside the bounds to outside the bounds of the control
UIControlEventTouchUpInside	Fingertip lifted off the screen while inside the bounds of the control
UIControlEventTouchUpOutside	Fingertip lifted off the screen while outside the bounds of the control (likely after dragging)
UIControlEventTouchCancel	Touch sequence canceled by external event or destruction of control instance
UIControlEventValueChanged	Value represented by the control, such as a slider, changed
UIControlEventEditingDidBegin	Text field became the first responder when a touch entered its bounds
UIControlEventEditingChanged	Text field value changed by user
UIControlEventEditingDidEnd	Text field lost first responder status when a touch was made outside its bounds
UIControlEventEditingDidEndOnExit	Text field lost focus and editing session ended
UIControlEventAllTouchEvents	All touch events for a control
UIControlEventAllEditingEvents	All editing events for a text field
UIControlEventApplicationReserved	A range of bit values reserved for the application and custom controls
UIControlEventSystemReserved	A range of bit values reserved for internal frameworks
UIControlEventAllEvents	All events, including UIControlEventSystemReserved events

To register an instance with a `UIControl` descendant, you can use the `addTarget:action:forControlEvents:` as follows:

```
//  In an imaginary object
- (void)setup
{
    UIButton *button = [UIButton buttonWithType:UIButtonTypeRoundedRect];
    [button addTarget:self action:@selector(doSomething:)
        forControlEvents:UIControlEventTouchUpInside];
    [self addSubview:button];
    button.center = self.center;
}

- (void)doSomething:(id)sender
{
    NSLog(@"doing something as a result of a touch up inside on my button.");
}
```

You can register the same message as a handler for multiple control events by specifying multiple events in the `forControlEvents:` parameter of `addTarget:action:forControlE vents::`

```
// Register self for touch-down and touch-up events for a UIButton *saveButton

- (void)registerSaveButtonEventHandlers
{
```

```
[saveButton addTarget:self action:@selector(saveButtonPressed:)
    forControlEvents:UIControlEventTouchDown];

[saveButton addTarget:self action:@selector(saveButtonReleased:)
    forControlEvents:UIControlEventTouchUpInside | UIControlEventTouchUpOutside];

}

- (void)saveButtonPressed:(id)sender
{
    NSLog(@"Save button pressed.");
}

- (void)saveButtonReleased:(id)sender
{
    NSLog(@"Save button released.");
}
```

Registering the same action for multiple control events is very convenient when you wish to handle multiple events with a single method. However, be careful of overly complex methods that can be broken down into smaller, event-specific methods.

Standard Control Types

One perk of Cocoa Touch is the collection of control classes in UIKit. A user interface control pattern includes one or more user interface elements that work together to solve an input problem. It's clear that Apple put a lot of thought into the design of its controls, and developers can easily take advantage of that work. This section will introduce you to the most common types of controls.

Buttons

The simplest controls are buttons. The UIButton class is the foundation for most buttons, including custom button subclasses. Buttons are used extensively in Cocoa Touch applications and can be customized through display attributes or with specialized drawing code.

Creating buttons

You can use the buttonWithType: method of UIButton to create buttons of several different types. Table 7-2 describes the various UIButtonType keys.

Table 7-2. UIButtonType keys

UIButtonType key	Description
UIButtonTypeCustom	No button style. Instead, a custom drawRect: method should be defined.
UIButtonTypeRoundedRect	A rectangular button with rounded corners and a centered title. Used for all general-purpose buttons.

UIButtonType key	Description
UIButtonTypeDetailDisclosure	A circular button with a centered chevron (>). Used to display details of an associated object or record.
UIButtonTypeInfoLight	A light-colored, circular button with an italicized lowercase "i" character. Used to flip a view over to display settings or additional information.
UIButtonTypeInfoDark	A dark-colored, circular button with an italicized lowercase "i" character. Used to flip a view over to display settings or additional information.
UIButtonTypeContactAdd	A rectangular button with rounded corners and a centered title. Used to display either a list of contacts or a form for adding a new contact.

Creating and customizing buttons is easy with the `UIButton` class. Custom subclasses of `UIButton` allow developers to use more attractive buttons in their applications. In this example, a special `UIButton` subclass, `PrettyButton`, assigns special images to be used for the two main button states: `UIControlStateNormal` and `UIControlStateHighlighted`. The supplied images are used as stretchable graphics by invoking the `stretchableImageWithLeftCapWidth:topCapHeight:` method of `UIImage`. Stretchable images use the concept of caps to lock a portion of the left, right, top, and bottom of a graphic as non-stretchable, while allowing the center area outside of the caps to grow as needed:

```
// PrettyButton.h

#import <Foundation/Foundation.h>

@interface PrettyButton : UIButton {
    UIImage *standardImg;
    UIImage *hoverImg;
}

@end

// PrettyButton.m

#import "PrettyButton.h"

@implementation PrettyButton

- (id)init
{
    if(self = [super init]){
        if(!standardImg){
            UIImage *image = [UIImage imageNamed:@"standard.png"];
            standardImg = [image stretchableImageWithLeftCapWidth:12
                                                    topCapHeight:12];
        }

        if(!hoverImg){
            UIImage *image = [UIImage imageNamed:@"hover.png"];
            hoverImg = [image stretchableImageWithLeftCapWidth:12
                                                 topCapHeight:12];
        }
```

```
            [self setBackgroundImage:standardImg forState:UIControlStateNormal];
            [self setBackgroundImage:hoverImg forState:UIControlStateHighlighted];
            [self setTitleColor:[UIColor colorWithRed:.208
                                              green:.318
                                               blue:.525
                                              alpha:1.0]
                       forState:UIControlStateNormal];
            [self setTitleColor:[UIColor whiteColor]
                       forState:UIControlStateHighlighted];
            self.titleLabel.font = [UIFont fontWithName:@"Helvetica-Bold"
                                                   size:15.0];
            self.titleLabel.textAlignment = UITextAlignmentCenter;
    }
    return self;
}
@end

// ButtonsViewController.h

#import <UIKit/UIKit.h>

@interface ButtonsViewController : UIViewController {
}

- (void)createStandardButton;
- (void)createPrettyButton;
- (void)standardButtonPressed:(id)sender;
- (void)prettyButtonPressed:(id)sender;

@end

// ButtonsViewController.m

#import "ButtonsViewController.h"
#import "PrettyButton.h"

@implementation ButtonsViewController

- (void)viewDidLoad
{
    [super viewDidLoad];
    [self createStandardButton];
    [self createPrettyButton];
}

#pragma mark Standard button and click handler
- (void)createStandardButton
{
    // Create a button with a rounded-rectangle style
    UIButton *standardButton = [UIButton buttonWithType:UIButtonTypeRoundedRect];
    float x = 30.0;
    float y = 30.0;
    float width = 120.0;
    float height = 40.0;
    // Create the frame that determines the size of the button
```

```
    CGRect frame = CGRectMake(x, y, width, height);
    standardButton.frame = frame;
    // Set the title of the button for the normal state
    [standardButton setTitle:@"Standard" forState:UIControlStateNormal];
    // Add self as a target
    [standardButton addTarget:self action:@selector(standardButtonPressed:)
            forControlEvents:UIControlEventTouchUpInside];
    // Set the button as a subview of my view
    [self.view addSubview:standardButton];
}

- (void)standardButtonPressed:(id)sender
{
    NSLog(@"Standard button pressed.");
}

#pragma mark Pretty button and click handler
- (void)createPrettyButton
{
    // Create an instance of a custom button subclass
    PrettyButton *prettyButton = [[PrettyButton alloc] init];
    float x = 170.0;
    float y = 30.0;
    float width = 120.0;
    float height = 40.0;
    // Create the frame that determines the size of the button
    CGRect frame = CGRectMake(x, y, width, height);
    prettyButton.frame = frame;
    // Set the title of the button for the normal state
    [prettyButton setTitle:@"Custom" forState:UIControlStateNormal];
    // Add self as a target
    [prettyButton addTarget:self action:@selector(prettyButtonPressed:)
            forControlEvents:UIControlEventTouchUpInside];
    // Set the button as a subview of my view
    [self.view addSubview:prettyButton];
    [prettyButton release];
}

- (void)prettyButtonPressed:(id)sender
{
    NSLog(@"Pretty button pressed.");
}

@end
```

Figure 7-8 shows the results of this application: a standard UIButton with rounded corners, and a custom UIButton subclass with PNG files as background images.

Info buttons

Many iPhone applications fit into the category of *lightweight utilities*. These apps tend to be very simple in their presentation of information and require very little interaction. The utility application Xcode template provided by the iPhone SDK is typically the foundation of such apps. One common feature of utilities is the presence of a small icon that displays the letter "i" and acts as a button to display more information about the application. Additional information is most often displayed after a "flip" animation that reverses the view. This pattern is well known because it is the default behavior of the utility application template, and because Apple uses it in official iPhone applications such as the Weather application, as shown in Figure 7-9.

Figure 7-8. A standard UIButton and a custom UIButton subclass with a custom background

Figure 7-9. The Weather application features an information button in the lower-right corner

You can add an info button to your views by passing `UIButtonTypeInfoLight` or `UIButtonTypeInfoDark` to the static `buttonWithType` method of the `UIButton` class:

```
UIButton *lightButton = [UIButton buttonWithType:UIButtonTypeInfoLight];
UIButton *darkButton = [UIButton buttonWithType:UIButtonTypeInfoDark];
```

You can wire an information button to flip a view using Core Animation. In this example, the class `FlipperView` represents a main view on the screen. It in turn contains two subviews. The `frontView` variable is a pointer to an instance of `FrontView`, a `UIView` subclass. The `backView` variable is a pointer to an instance of `BackView`, also a `UIView` subclass. When the `toggle` method is called, an animation transaction is created and a transition is assigned to the `FlipperView` instance:

```
// FrontView.h

#import <UIKit/UIKit.h>

@interface FrontView : UIImageView {
}

@end

#import "FrontView.h"
#import "FlipperView.h"

@implementation FrontView

#define PADDING 10.0

- (id)initWithFrame:(CGRect)frame
{
    if (self = [super initWithFrame:frame]) {
        infoButton = [UIButton buttonWithType:UIButtonTypeInfoDark];
        [infoButton addTarget:self action:@selector(showInfo:)
            forControlEvents:UIControlEventTouchUpInside];
        infoButton.center = CGPointMake(
            self.frame.size.width - infoButton.frame.size.width/2 - PADDING,
            self.frame.size.height - infoButton.frame.size.height/2 - PADDING
        );
        [self addSubview:infoButton];
        self.backgroundColor = [UIColor clearColor];
        self.image = [UIImage imageNamed:@"front.png"];
        self.userInteractionEnabled = YES;
    }
    return self;
}

- (void)showInfo:(id)sender
{
    [(FlipperView *)self.superview toggle];
}

@end
```

```
// FlipperView.h

#import <UIKit/UIKit.h>

@class FrontView;
@class BackView;

@interface FlipperView : UIView {
    BOOL isFlipped;
    FrontView *frontView;
    BackView *backView;
}

@property (assign) BOOL isFlipped;

- (void)toggle;

@end

// FlipperView.m

#import "FlipperView.h"
#import "FrontView.h"
#import "BackView.h"

@implementation FlipperView

@synthesize isFlipped;

- (id)initWithFrame:(CGRect)frame
{
    if(self = [super initWithFrame:frame]){
        frontView = [[FrontView alloc] initWithFrame:self.frame];
        backView = [[BackView alloc] initWithFrame:self.frame];

        //    Add the front view as the main content to start off.
        [self addSubview:frontView];

        //    Insert the back view under the front view, hidden.
        [self insertSubview:backView belowSubview:frontView];

        self.backgroundColor = [UIColor clearColor];
        self.clipsToBounds = YES;
    }
    return self;
}

- (void)toggle
{
    [UIView beginAnimations:nil context:NULL];
    [UIView setAnimationDuration:1];

    UIViewAnimationTransition direction;

    if(isFlipped){
        direction = UIViewAnimationTransitionFlipFromLeft;
```

```
        isFlipped = NO;
    }else{
        direction = UIViewAnimationTransitionFlipFromRight;
        isFlipped = YES;
    }

    //    Mid-animation, swap the views.
    [self exchangeSubviewAtIndex:0 withSubviewAtIndex:1];

    [UIView setAnimationTransition:direction forView:self cache:YES];
    [UIView commitAnimations];
}

- (void)dealloc
{
    [frontView release];
    [backView release];
    [super dealloc];
}


@end
```

You should be careful having an information button trigger views outside of the flip, though you needn't necessarily avoid it. As with most user experience programming for Cocoa Touch, it's best to stick to the known paths and perfect the nuances, but at the same time you shouldn't hesitate to evaluate new patterns.

Modal Buttons

Modal controls are UI controls that change state across two or more steps as part of a single expression of user intent. A good example of a modal control is the "Buy Now/ Install/Installed" button in the mobile App Store application. You install applications from the App Store on the mobile device by tapping the button several times at specific points in the sequence.

Modal controls are very effective for operations that should be confirmed by users. They can also be excellent triggers when one of a set of options is available, depending on context. For example, the individual record screen in the Contacts application allows you to click an Edit button that changes the view to an editable view of the record. The Edit button then becomes a Done button that can be used to cancel or commit any editing actions that have taken place. Figure 7-10 shows the modal control in the Contacts application.

Figure 7-10. The Contacts application uses a modal control for "Edit" and "Done"

Providing secondary confirmations is a common and useful pattern for preventing destructive operations such as deletions.

Creating a modal button subclass

This example illustrates just one way to create a modal button. There are several options, including building custom `UIView` objects from scratch. I chose to take a simpler route and simply subclassed `UIButton` in order to keep the interface consistent, reduce duplication of code, and avoid tiny deviations in the look and feel of my button and standard `UIButton` instances. The implementation is fairly standard. `ModalButton` is a subclass of `UIButton` and overloads the `touchesEnded:withEvent:` method to add custom handling behavior. Specifically, each touch deactivates the button and sets an `NSTimer` to fire after a delay, presenting the next state in a finite sequence of modes. The delay emulates what might happen if you were to design the App Store button that triggers an application download over HTTP and acts as an implicit opt-in agreement to purchase an application (versus a one-touch sequence, which might be triggered by accident).

The `ModalButton` class adheres to the `UIControl` target-action messaging mechanism. In this example, the main view controller, `ModalButtonViewController`, adds itself as a target for the `UIControlEventTouchUpInside` control event. Doing so sends a specific action message to the `ModalButtonViewController` instance when a `UITouch` sequence begins and ends inside the bounds of the `ModalButton`:

```
#import "ModalButtonViewController.h"

@implementation ModalButtonViewController

- (void)viewDidLoad
{
    [super viewDidLoad];
    button = [[ModalButton alloc] init];
    [button addTarget:self action:@selector(performModalActionForButton:)
        forControlEvents:UIControlEventTouchUpInside];
    [button retain];
    [self.view addSubview:button];
    button.center = CGPointMake(160.0, 210.0);
}

- (void)performModalActionForButton:(id)sender
{
    ModalButton *btn = (ModalButton *)sender;
    ModalButtonMode mode = btn.mode;
    NSLog(@"The button mode is: %d", mode);
}

- (void)dealloc
{
    [button release];
    [super dealloc];
}

@end
```

The definition of the ModalButton class follows:

```
#import <UIKit/UIKit.h>

static NSString *download = @"Download";
static NSString *downloading = @"Downloading...";
static NSString *install = @"Install";
static NSString *installing = @"Installing...";
static NSString *complete = @"Complete!";

static UIImage *heart = nil;
static UIImage *clover = nil;

typedef enum {
    ModalButtonModeDefault = 0,
    ModalButtonModeDownload,
    ModalButtonModeInstall,
    ModalButtonModeComplete,
} ModalButtonMode;


@interface ModalButton : UIButton {
    ModalButtonMode mode;
    NSTimer *timer;
    UIActivityIndicatorView *indicator;
}
```

```objc
@property (readonly) ModalButtonMode mode;

- (void)update:(NSTimer *)theTimer;

@end

#import "ModalButton.h"

@interface ModalButton (PrivateMethods)
- (void)handleTap;
@end

@implementation ModalButton

@synthesize mode;

- (id) init
{
    self = [self initWithFrame:CGRectMake(0.0, 0.0, 118.0, 118.0)];
    return self;
}

- (id)initWithFrame:(CGRect)frame
{
    if (self = [super initWithFrame:frame]) {
        mode = ModalButtonModeDefault;
        heart = [UIImage imageNamed:@"heart.png"];
        clover = [UIImage imageNamed:@"clover.png"];

        indicator = [[UIActivityIndicatorView alloc]
            initWithActivityIndicatorStyle:UIActivityIndicatorViewStyleGray];
        indicator.hidesWhenStopped = YES;
        [self addSubview:indicator];
        [self bringSubviewToFront:indicator];
        indicator.center = self.center;

        self.backgroundColor = [UIColor clearColor];
        [self setTitleColor:[UIColor darkGrayColor] forState:UIControlStateNormal];
        [self setTitleColor:[UIColor whiteColor] forState:UIControlStateDisabled];
        self.titleEdgeInsets = UIEdgeInsetsMake(-90.0, 0, 0, 0);
        self.font = [UIFont fontWithName:@"Helvetica" size:12.0f];
        [self setBackgroundImage:heart forState:UIControlStateNormal];

        [self update:nil];
    }
    return self;
}

- (void)touchesEnded:(NSSet *)touches withEvent:(UIEvent *)event
{
    UITouch *touch = [touches anyObject];
    if([self hitTest:[touch locationInView:self] withEvent:event]){
        [self handleTap];
    }
```

```
        [super touchesEnded:touches withEvent:event];
}

- (void)handleTap
{
    self.enabled = NO;
    NSString *title;
    switch(mode){
        case ModalButtonModeDownload:
            title = downloading;
            break;
        case ModalButtonModeInstall:
            title = installing;
            break;
        default:
            break;
    }
    [self setTitle:title forState:UIControlStateNormal];

    if([timer isValid]){
        [timer invalidate];
    }
    timer = [NSTimer timerWithTimeInterval:5.0 target:self
        selector:@selector(update:)
        userInfo:nil repeats:NO];
    [timer retain];
    [[NSRunLoop currentRunLoop] addTimer:timer forMode:NSDefaultRunLoopMode];

    [indicator startAnimating];
}

- (void)update:(NSTimer *)theTimer
{
    NSString *title;
    // Toggle mode
    switch(mode){
        case ModalButtonModeDefault:
            mode = ModalButtonModeDownload;
            title = download;
            self.enabled = YES;
            break;
        case ModalButtonModeDownload:
            mode = ModalButtonModeInstall;
            title = install;
            self.enabled = YES;
            break;
        case ModalButtonModeInstall:
            mode = ModalButtonModeComplete;
            title = complete;
            [self setBackgroundImage:clover forState:UIControlStateNormal];
            self.enabled = NO;
            break;
        default:
            self.enabled = NO;
            return;
```

```
    }
    [self setTitle:title forState:UIControlStateNormal];

    if([timer isValid]){
        [timer invalidate];
    }
    [indicator stopAnimating];
}

- (void)dealloc
{
    if([timer isValid]){
        [timer invalidate];
    }
    [timer release];
    [indicator release];
    [super dealloc];
}
@end
```

Sliders

Sliders are used to modify a value over a finite range. A slider is made up of two parts: a *thumb* and a *track*. The thumb is draggable across the width of the slider. The track defines the width of the slider and gives a proportional context for the position of the thumb. The standard slider class included in UIKit is UISlider.

You can create a slider by adding an instance as a subview to your window or base UIView. As with other UIControl classes, you should use the target-action mechanism to set at least one target for events generated by the slider. Your slider can send action messages to associated targets either as its value changes or upon release of the slider. The UIControlEvent type for a changed value in a slider is UIControlEventValue Changed. It's possible for a user to change values rapidly and often, so developers should keep expensive operations to a minimum:

```
// Code for setting up a UISlider

#import "SliderViewController.h"

@implementation SliderViewController

- (void)viewDidLoad
{
    standardSlider = [[UISlider alloc] initWithFrame:CGRectMake(60,20,200,40)];
    [self.view addSubview:standardSlider];
    [standardSlider addTarget:self action:@selector(sliderChanged:)
        forControlEvents:UIControlEventValueChanged];
    [super viewDidLoad];
}

- (void)sliderChanged:(id)sender
{
    UISlider *slider = (UISlider *)sender;
```

```
        NSLog(@"Slider changed. New value: %f.", slider.value);
}

- (void)dealloc
{
    [standardSlider release];
    [super dealloc];
}

@end
```

Figure 7-11 shows a standard UISlider.

Figure 7-11. A standard UISlider control

You can customize the look of a UISlider by changing the images representing the thumb and by manipulating the color of the track bar:

```
// Code for overriding the thumb image and track

#import "SliderViewController.h"

@implementation SliderViewController

- (void)viewDidLoad
{
    standardSlider = [[UISlider alloc] initWithFrame:CGRectMake(60,20,200,40)];
    [self.view addSubview:standardSlider];

    customSlider = [[UISlider alloc] initWithFrame:CGRectMake(60,60,200,40)];
    customSlider.backgroundColor = [UIColor clearColor];
```

```
    [customSlider setThumbImage:[UIImage imageNamed:@"thumbOff.png"]
        forState:UIControlStateNormal];
    [customSlider setThumbImage:[UIImage imageNamed:@"thumbOn.png"]
        forState:UIControlStateHighlighted];

    UIImage *leftTrack = [[UIImage imageNamed:@"leftTrack.png"]
        stretchableImageWithLeftCapWidth:5.0 topCapHeight:0.0];
    UIImage *rightTrack = [[UIImage imageNamed:@"rightTrack.png"]
        stretchableImageWithLeftCapWidth:5.0 topCapHeight:0.0];

    [customSlider setMinimumTrackImage:leftTrack forState:UIControlStateNormal];
    [customSlider setMaximumTrackImage:rightTrack forState:UIControlStateNormal];

    [self.view addSubview:customSlider];

    [super viewDidLoad];
}

- (void)dealloc
{
    [standardSlider release];
    [customSlider release];
    [super dealloc];
}

@end
```

Figure 7-12 shows a UISlider subclass with a custom thumb image.

Figure 7-12. A customized UISlider control

Tables and Pickers

The UITableView and UIPickerView classes display lists of values to users. Neither class is a descendant of UIControl, but both are useful as input interfaces. You can use cells in tables to show a list of options that accepts multiple concurrent choices. A picker view is a good option for showing a list that accepts only one selection at a time.

Tables are often used to display lists of items with a selectable cell for each possible value. Tables are excellent options for displaying long, one-dimensional lists or lists with groups of smaller, related lists. You should consider using a UITableView when the values in your dataset aren't known or intuitive to users. Finite ordered sets that users recognize, such as the names of months or hours in the day, can be displayed using controls that require less screen area because users can easily infer the values that are offscreen based on the visible at any moment elements. Tables can display items from a large set, giving more context and allowing faster scrolling and indexing (using an index ribbon). The Ringtones screen in the Contacts application uses a standard UITableView with groups of selectable cells, as shown in Figure 7-13.

Figure 7-13. UITableView on the Ringtones screen of the Contacts application

Like UITableView, UIPickerView is not a descendant of UIControl but acts as a user input control. You've seen pickers used in the Clock, Calendar, and Mail applications, among others. Most UIKit controls are metaphors for real-world interfaces. Buttons, sliders, and switches all mimic the behavior of a physical interface element. The hardware version of a UIPickerView is a group of vertically spinning dials reminiscent of a classic slot machine interface.

Figure 7-14 shows a simple picker from the `UICatalog` application included in the iPhone SDK examples.

Figure 7-14. Simple UIPickerView from the UICatalog example application

Unlike `UITableView` instances, pickers can display multidimensional data as vertically parallel lists. For example, consider the `UIDatePicker` class included in `UIKit`. Users can choose date components individually when setting a date in a `UIDatePicker`. Depending on the mode of operation assigned to a date picker instance, the interface will display one of the following sets of components:

- For `UIDatePickerModeTime`, the picker will display hours, minutes, and (optionally) an AM/PM selection component. The order of the components depends on global localization contexts.

- For `UIDatePickerModeDate`, the picker will display months, days of the month, and years.

- For `UIDatePickerModeDateAndTime`, the picker displays dates (in one component), hours, minutes, and (optionally) an AM/PM selection component.

- For `UIDatePickerModeCountdownTimer`, the picker displays hour and minute values representing an interval for a countdown.

You can retrieve the mode for a `UIDatePicker` instance using its `datePickerMode` property.

When working with a `UIPickerView`, you should follow the conventions outlined by Apple for displaying values. You should avoid listing units in a value component except when the unit type is the value. For example, the AM/PM selection component displayed by the `UIDatePicker` class is appropriate because both "AM" and "PM" are values. In cases where you'd like to display a set of non-value units for a component, you can use a non-standard approach to build the display. There are safer methods, but this approach provides a quick proof of concept:

```
#import "LabeledPickerView.h"

@interface _UIPickerViewSelectionBar : UIView
@end

@implementation LabeledPickerView

- (void)drawRect:(CGRect)rect
{
    // Draw the label if it needs to be drawn
    if(fixedLabel == nil){
        fixedLabel = [[UILabel alloc]
            initWithFrame:CGRectMake(0.0, 0.0, 200.0, 40.0)];
        fixedLabel.font = [UIFont boldSystemFontOfSize:20.0];
        fixedLabel.textColor = [UIColor colorWithRed:0 green:0 blue:0 alpha:1];
        fixedLabel.backgroundColor = [UIColor clearColor];
        fixedLabel.shadowColor = [UIColor whiteColor];
        fixedLabel.shadowOffset = CGSizeMake(0, 1);
        fixedLabel.text = @"...is tasty.";
        NSArray *svs = [self subviews];
        UIView *v;
        for(v in svs){
            if([v isKindOfClass:[_UIPickerViewSelectionBar class]]){
                CGPoint c = CGPointMake(v.center.x + 40.0, v.center.y - 86);
                fixedLabel.center = c;
                [v addSubview:fixedLabel];
                return;
            }
        }
    }
}

- (void)dealloc
{
    [fixedLabel release];
    [super dealloc];
}

@end
```

The key operation is in the search for an instance of _UIPickerViewSelectionBar and the addition of a custom UILabel instance as a subview to the selection bar. The result can be seen in Figure 7-15.

Figure 7-15. Adding a custom label to the UIPickerView selection bar

Search Bars

Search bars gather user input and allow applications to respond to that input in real time. A simple example is using a search bar at the top of the screen to filter a list of records in a table view based on string pattern matching. To use a search bar, you must create an instance of UISearchBar and add it to your view. You must also set the delegate property of the UISearchBar instance to an object that implements the UISearchBarDelegate protocol. Each method in the UISearchBarDelegate protocol handles a different user action. For example, you can implement the searchBar:textDid Change: method to develop progressive searches (a.k.a., auto-complete searches), or you can choose to execute searches only when editing ends using the searchBarSearch ButtonClicked: and searchBarTextDidEndEditing: methods.

The TileSearchViewController class is from an example application that filters colored tiles based on color names input by a user:

```objc
#import "TileSearchViewController.h"
#import "Board.h"

@implementation TileSearchViewController

- (void)viewDidLoad
{
    searchBar = [[UISearchBar alloc]
        initWithFrame:CGRectMake(0.0, 0.0, 320.0, 44.0)];
    searchBar.autocorrectionType = UITextAutocorrectionTypeNo;
    searchBar.autocapitalizationType = UITextAutocapitalizationTypeNone;
    searchBar.showsCancelButton = NO;

    //    Set myself as the delegate of the search bar.
    searchBar.delegate = self;

    [self.view addSubview:searchBar];
    if(board == nil) board = [[Board alloc]
        initWithFrame:CGRectMake(0.0, 45.0, 320.0, 415.0)];
    [self.view addSubview:board];
}


#pragma mark UISearchBarDelegate methods
- (void)searchBarTextDidBeginEditing:(UISearchBar *)theSearchBar
{
    //    Show cancel button while editing
    searchBar.showsCancelButton = YES;
}

- (void)searchBarTextDidEndEditing:(UISearchBar *)theSearchBar
{
    //    Hide cancel button when editing ends
    searchBar.showsCancelButton = NO;
}

- (void)searchBar:(UISearchBar *)theSearchBar
    textDidChange:(NSString *)searchText
{
    [board filterForColorName:searchText];
}

- (void)searchBarCancelButtonClicked:(UISearchBar *)theSearchBar
{
    //    Repopulate the last color unless the user cleared the search
    if (theSearchBar.text.length > 0){
        searchBar.text = lastSearch;
    }else{
        searchBar.text = @"";
    }

    [searchBar resignFirstResponder];
}

- (void)searchBarSearchButtonClicked:(UISearchBar *)theSearchBar
```

```
{
    [lastSearch release];
    lastSearch = [theSearchBar.text copy];
    [searchBar resignFirstResponder];
}


- (void)dealloc
{
    [searchBar release];
    [board release];
    [lastSearch release];
    [super dealloc];
}

@end
```

Figure 7-16 shows a screenshot of the TileSearch example application.

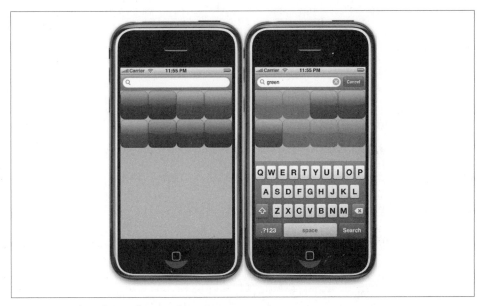

Figure 7-16. The TileSearch example application

Segmented Controls

Segmented controls provide a compact, persistent grouping of buttons that switch between views. According to the mobile HIG, segmented controls should provide feedback to users by swapping views or otherwise appropriately updating the UI. The feedback should be immediate, avoiding animation effects. It's conceivable to use segmented controls for complex view management. For example, the following code illustrates the use of segmented controls to manage pagination for a UIScrollView with fixed, known content. This approach isn't standard for pagination, but it may prove to be more usable than the UIPageControl class, which uses very tiny buttons for paging:

```objc
#import "RootViewController.h"
#import "SegmentedPaginatorAppDelegate.h"

@implementation RootViewController

- (void)viewDidLoad
{
    UISegmentedControl *segmentedControl = [[[UISegmentedControl alloc]
        initWithItems:
                                            [NSArray arrayWithObjects:
                                            @"1",
                                            @"2",
                                            @"3",
                                            @"4",
                                            @"5",
                                            nil]] autorelease];
    [segmentedControl addTarget:self action:@selector(segmentChosen:)
        forControlEvents:UIControlEventValueChanged];
    segmentedControl.frame = CGRectMake(0, 0, 200, 30.0);
    segmentedControl.segmentedControlStyle = UISegmentedControlStyleBar;
    segmentedControl.momentary = NO;

    self.navigationItem.titleView = segmentedControl;

    UIScrollView *scrollView = [[UIScrollView alloc]
        initWithFrame:self.view.frame];
    scrollView.contentSize = CGSizeMake(320.0, 422.0);
    scrollView.scrollEnabled = NO;
    [scrollView setPagingEnabled:YES];

    UIView *viewOne = [[UIView alloc]
        initWithFrame:CGRectMake(0.0, 0.0, 320.0, 422.0)];
    viewOne.backgroundColor = [UIColor redColor];
    UIView *viewTwo = [[UIView alloc]
        initWithFrame:CGRectMake(320.0, 0.0, 320.0, 422.0)];
    viewTwo.backgroundColor = [UIColor blueColor];
    UIView *viewThree = [[UIView alloc]
        initWithFrame:CGRectMake(640.0, 0.0, 320.0, 422.0)];
    viewThree.backgroundColor = [UIColor greenColor];
    UIView *viewFour = [[UIView alloc]
        initWithFrame:CGRectMake(960.0, 0.0, 320.0, 422.0)];
    viewFour.backgroundColor = [UIColor orangeColor];
    UIView *viewFive = [[UIView alloc]
        initWithFrame:CGRectMake(1280.0, 0.0, 320.0, 422.0)];
    viewFive.backgroundColor = [UIColor yellowColor];

    [scrollView addSubview:viewOne];
    [scrollView addSubview:viewTwo];
    [scrollView addSubview:viewThree];
    [scrollView addSubview:viewFour];
    [scrollView addSubview:viewFive];
    scrollView.contentSize = CGSizeMake(1600.0, 422.0);
```

```
    [viewOne release];
    [viewTwo release];
    [viewThree release];
    [viewFour release];
    [viewFive release];

    self.view = scrollView;
}

- (void)segmentChosen:(id)sender
{
    UISegmentedControl* segmentedControl = sender;
    NSUInteger i = [segmentedControl selectedSegmentIndex];
    UIScrollView *scrollView = (UIScrollView *)self.view;
    [scrollView
        scrollRectToVisible:CGRectMake((320.0 * i), 0.0, 320.0, 422.0)
        animated:YES];
}

@end
```

Figure 7-17 shows the example running in the iPhone emulator.

Figure 7-17. Example use of segmented control for pagination

Obviously, this implementation is just a simple example. It wouldn't be terribly difficult to develop a more usable clone of the `UIPageControl` element based on the `UISegmentedControl` class and the segmented control pattern, though such an exercise is outside the scope of this book.

Overall, working with segmented controls is simple and adds a rich supplement to your organization and navigation toolkit.

Scrolling Controls

The iPhone handles scrolling in a very intuitive manner. The timing of animations and the touch and gesture recognition build muscle memory and acclimate users to the idea of UI elements moving on- and offscreen. Apple sets some precedents for the timing of scrolling, zooming, and panning, and developers who wish to provide a consistent user experience should build their visual transitions around those precedents.

You can use the `UIScrollView` class to create views that scroll horizontally, vertically, or both. Scrollviews respond to touches, drags, and flick gestures. Dragging a finger across the screen slowly scrolls the content in the scrolling view in parallel. Quickly flicking across the screen will throw the content in the direction of the swipe, and it will continue moving after the touch sequence has ended. Adding a scrolling view to an application is no different from adding any other view to the screen because `UIScrollView` is a subclass of `UIView`. Developers can enable a bounce effect using a property of the `UIScrollView` instance called `bounces`.

The Photos application uses a scrolling view to display full-screen photographs sequentially. Scrolling views can scroll incrementally or in larger units that correspond to the width of the screen, also known as *paging*. You can enable paging with the `pagingEnabled` attribute of `UIScrollView`.

The following example creates a basic `UIScrollView` subclass for displaying a sequence of full-screen images. Bouncing is enabled, as is paging. You can enable or disable a subtle scrollbar for scrolling views using the `showsHorizontalScrollIndicator` and `showsVerticalScrollIndicator` properties. Small details like scrollbars can make a big impact for users. In some cases, such as a photo gallery, even a subtle scrollbar can be distracting. In other cases, such as the current step in a process or remaining copy in an article, a scrollbar provides important context. `UIKit` classes offer a great deal of configurability and allow design and development teams to create the most appropriate interface for the task at hand:

```
// GalleryView.h

#import <UIKit/UIKit.h>

@interface GalleryView : UIScrollView {
}

- (void)addImage:(UIImage *)image;
```

```
@end

// GalleryView.m

#import "GalleryView.h"

@implementation GalleryView

- (id)initWithFrame:(CGRect)frame
{
    if(self = [super initWithFrame:frame]){
        self.backgroundColor = [UIColor blackColor];
        self.scrollEnabled = YES;
        self.pagingEnabled = YES;
        self.bounces = YES;
        self.directionalLockEnabled = NO;
    }
    return self;
}

- (void)addImage:(UIImage *)image
{
    int imageCount = [self.subviews count];
    float newContentWidth = ((float)imageCount + 1.0) * 320.0;
    CGSize newContentSize = CGSizeMake(newContentWidth, 460.0);
    UIImageView *imageView = [[UIImageView alloc]
        initWithFrame:CGRectMake((imageCount * 320.0), 0.0, 320.0, 460.0)];

    self.contentSize = newContentSize;

    imageView.image = image;
    [self addSubview:imageView];
    [imageView release];
}

@end

// ImageGalleryViewController.m

#import "ImageGalleryViewController.h"

@implementation ImageGalleryViewController

- (void)viewDidLoad
{
    GalleryView *galleryView = [[GalleryView alloc]
                    initWithFrame:[UIScreen mainScreen].applicationFrame];

    [galleryView addImage:[UIImage imageNamed:@"murray.png"]];
    [galleryView addImage:[UIImage imageNamed:@"murray.png"]];
    [galleryView addImage:[UIImage imageNamed:@"murray.png"]];
    [galleryView addImage:[UIImage imageNamed:@"murray.png"]];
    [galleryView addImage:[UIImage imageNamed:@"murray.png"]];

    self.view = galleryView;
```

```
    [galleryView release];

    [super viewDidLoad];
}

@end
```

Scrolling views can also be used for subtle effects. The following example shows the development of a custom UIControl subclass that uses a UIScrollView to create scrolling interaction. It is possible to develop a fully custom scrolling interface without the use of a UIScrollView, but the available UIKit classes help provide consistency in subtle ways, such as in the timing of animations:

```
// ScrollingControlViewController.h

#import <UIKit/UIKit.h>

@class Scroller;

@interface ScrollingControlViewController : UIViewController {
    Scroller *scroller;
}

- (void)scrollerDidScroll:(id)sender;

@end

// ScrollingControlViewController.m

#import "ScrollingControlViewController.h"
#import "Scroller.h"

@implementation ScrollingControlViewController

- (void)viewDidLoad
{
    CGRect f = CGRectMake(0.0, 0.0, 320.0, 460.0);
    UIImageView *backgroundView = [[[UIImageView alloc]
        initWithFrame:f]
        autorelease];
    backgroundView.image = [UIImage imageNamed:@"background.png"];
    [self.view addSubview:backgroundView];

    scroller = [[Scroller alloc]
        initWithFrame:CGRectMake(0.0, 100.0, 320.0, 60.0)];
    [scroller addTarget:self action:@selector(scrollerDidScroll:)
        forControlEvents:UIControlEventApplicationReserved];
    [self.view addSubview:scroller];

    f.size.height = 126.0;
    UIImageView *topView = [[[UIImageView alloc] initWithFrame:f]
        autorelease];
    topView.image = [UIImage imageNamed:@"top.png"];
    [self.view addSubview:topView];
    [super viewDidLoad];
```

```objc
}

- (void)scrollerDidScroll:(id)sender
{
    NSLog(@"Scroller did scroll.");
}

- (void)dealloc
{
    [scroller release];
    [super dealloc];
}

@end
```

```objc
// Scroller.h

#import <UIKit/UIKit.h>

@interface Scroller : UIControl <UIScrollViewDelegate> {
    UIScrollView *numbersView;
    CGPoint touchPoint;
    CGPoint scrollerPoint;
}

@end
```

```objc
// Scroller.m

#import "Scroller.h"
#import "NumView.h"

@interface Scroller (PrivateMethods)

- (void)snapToClosestNumber;

@end

@implementation Scroller

#define WIDTH 500.0
#define NUM 10
#define NUM_WIDTH (WIDTH / NUM)
#define HALF (NUM_WIDTH / 2)
#define HEIGHT 40.0
#define INSET_WIDTH 160.0

- (void)snapToClosestNumber
{
    CGPoint coff = numbersView.contentOffset;
    float normalizedX = coff.x + INSET_WIDTH;
    double diff = fmod(normalizedX, NUM_WIDTH);

    // Move to the left or right, as needed
    if(diff < NUM_WIDTH){
```

```
        // If we're at the max...
        if(normalizedX == WIDTH){
            normalizedX -= NUM_WIDTH;
        }
        normalizedX -= diff;
    }else{
        normalizedX += diff;
    }

    float leftX = normalizedX - INSET_WIDTH + HALF;

    [numbersView scrollRectToVisible:CGRectMake(leftX, 0.0, 320.0, HEIGHT)
        animated:YES];
}

- (void)scrollViewDidScroll:(UIScrollView *)scrollView
{
    NSLog(@"sending actions for UIControlEventApplicationReserved.");
    [self sendActionsForControlEvents:UIControlEventApplicationReserved];
}

- (void)scrollViewDidEndDragging:(UIScrollView *)scrollView
        willDecelerate:(BOOL)decelerate
{
    [self snapToClosestNumber];
}

- (void)scrollViewDidEndDecelerating:(UIScrollView *)scrollView
{
    [self snapToClosestNumber];
}

- (id)initWithFrame:(CGRect)frame
{
    if (self = [super initWithFrame:frame]) {
        numbersView = [[UIScrollView alloc]
            initWithFrame:CGRectMake(0.0, 0.0, 320.0, 66.0)];
        numbersView.delegate = self;
        numbersView.showsHorizontalScrollIndicator = NO;
        numbersView.showsVerticalScrollIndicator = NO;
        numbersView.delaysContentTouches = NO;
        numbersView.bounces = YES;

        self.backgroundColor = [UIColor clearColor];
        //    Add in a bunch of subs
        NSUInteger i = 0;
        NumView *numView;
        CGRect frame;
        for(i; i < NUM; i++){
            frame = CGRectMake((i * NUM_WIDTH), 20.0, NUM_WIDTH, HEIGHT);
            numView = [[[NumView alloc] initWithFrame:frame number:i] autorelease];
            [numbersView addSubview:numView];
            numView.frame = frame;
        }
        [self addSubview:numbersView];
```

```
        numbersView.contentSize = CGSizeMake(WIDTH, HEIGHT);
        numbersView.contentInset = UIEdgeInsetsMake(0.0,
                                                              INSET_WIDTH,
                                                              0.0,
                                                              INSET_WIDTH);

    }
    return self;
}

- (void)dealloc
{
    [numbersView release];
    [super dealloc];
}


@end

// NumView.h

#import <UIKit/UIKit.h>

@interface NumView : UILabel {
}

- (id)initWithFrame:(CGRect)frame number:(NSUInteger)num;

@end

// NumView.m

#import "NumView.h"

@implementation NumView

- (id)initWithFrame:(CGRect)frame number:(NSUInteger)num
{
    if (self = [super initWithFrame:frame]) {
        self.text = [NSString stringWithFormat:@"%d", num];
        self.backgroundColor = [UIColor clearColor];
        self.textColor = [UIColor whiteColor];
        self.textAlignment = UITextAlignmentCenter;
        self.shadowColor = [UIColor darkGrayColor];
        self.shadowOffset = CGSizeMake(0, 1.0);
        self.font = [UIFont boldSystemFontOfSize:36.0];
    }
    return self;
}

@end
```

Figure 7-18 shows the custom control.

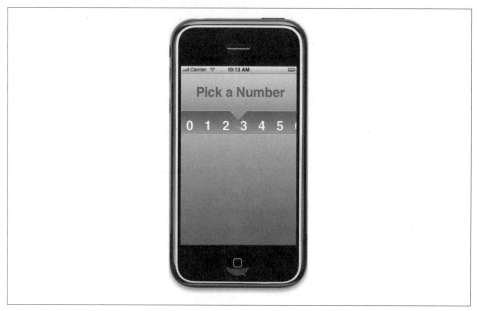

Figure 7-18. Custom scrolling UIControl subclass

Tables and Embedded Controls

A table in Cocoa Touch represents a vertical list of items, with each item assigned to a cell or row in the table. If the list of items is multi-dimensional, such as a list of email messages in the Mail application, each table cell should display a succinct summary, label, or other high-level indication of the object assigned to the cell. Users can get additional information for each row by tapping the row. When a user taps the row, an event is sent to the delegate of the table, which is an object that conforms to the UITableViewDelegate protocol. This protocol defines methods for handling interaction with cells. A table delegate defines the methods it wishes to handle, and the table controller automatically handles the communication between the user interface and the delegate. (The availability of free implementations of design patterns like the delegate pattern is one of the benefits that Cocoa Touch controller classes offer developers.)

You can embed UIControl instances in table cells to add functionality within the context of a single row in your dataset. For example, users will often need to delete a record using a button, as with the Mail application, or change the order of rows in a table. The table delegate protocol, UITableViewDelegate, allows developers to handle user-initiated edits like reordering, deletion, and insertion of new rows into the table.

The controls you use in table cells should be chosen with attention to the interactive nature of the tables and cells themselves. Most tables are embedded in a scrolling view that responds to gestures with movement along the y-axis. In those cases, it would be difficult to present users with an embedded control that also responds to gestures along the y-axis. Tables that act as views for navigation controllers and allow users to drill down through a hierarchical dataset already handle taps.

Adding controls to table cells is simple. There are three approaches to developing custom table cells:

- Create an instance of UITableViewCell and set its public properties, using the default object and layout with custom content.
- Create an instance of UITableViewCell and customize it using public properties and methods for managing subviews. This can include adding subviews to the content View of the cell, setting the accessoryView appropriately, adding an icon or image to the left side of the cell using the imageView property, and manipulating the visual characteristics such as background color and text properties. In most cases, a standard UITableViewCell is an appropriate starting point because it provides customized consistency in the user experience.
- Subclass UITableViewCell with overrides for initialization, layout, and event handling, and add any functionality desired.

The simplest option is to create an instance of UITableViewCell and customize it using properties. Working with standard cell properties minimizes the amount of code required for customization, which in turn limits the risk of bugs. Users benefit from standard cells because familiarity speeds the learning curve for new users.

Adding subviews to a standard UITableViewCell is slightly more complex. Developers must manage the layout of subviews and support resizing and repositioning subviews to support rotation. In exchange, more customization is possible.

The most complex and flexible option is to subclass UITableViewCell. Developers can override the core functionality of a cell to render complex artwork or manage touch events in novel ways.

Standard table view cells include three subviews that display content to users. On the left side of each cell is a view called imageView. The center and majority of the cell is occupied by a view that displays the main cell content, sensibly called the content View. To the right of the contentView is a subview that can display standard indicators such as a checkmark, chevron, custom graphic, and even controls. The view for accessories is called the accessoryView.

Passive Indicators

The disclosure indicator is a familiar control for diving deeper into a stack of UINavigationController instances from within a UITableView. The disclosure indicator looks like a chevron and shows that more information is available for a row. You can set the indicator type for a cell by setting the `accessoryType` property of the `UITable Cell` instance to `UITableViewCellAccessoryDisclosureIndicator`:

```
- (UITableViewCell *)tableView:(UITableView *)tableView
        cellForRowAtIndexPath:(NSIndexPath *)indexPath
    {
        static NSString *CellIdentifier = @"OMG_Its_a_Cell";

        UITableViewCell *cell = [tableView
            dequeueReusableCellWithIdentifier:CellIdentifier];
        if (cell == nil) {
            cell = [[[UITableViewCell alloc] initWithFrame:CGRectZero
                reuseIdentifier:CellIdentifier] autorelease];
            cell.accessoryType = UITableViewCellAccessoryDisclosureIndicator;
        }

        cell.text = [nodes objectAtIndex:indexPath.row];

        return cell;
    }
```

You can display a checkmark as an accessory by assigning the `UITableViewCellAc cessoryCheckmark` constant to `accessoryType`:

```
cell.accessoryType = UITableViewCellAccessoryCheckmark;
```

Assigning a custom image to an accessory view is nearly as simple:

```
cell.accessoryView = [[[UIImageView alloc] initWithImage:myImage] autorelease];
```

Active Indicators and Control Accessories

Developers can assign any `UIView` to the `accessoryView` property, including controls like sliders, buttons, and switches.

If the additional detail for the object represented by the cell consists of configuration options, or if the cell has multiple subviews that track touches independently, you can use a detail disclosure button by assigning `UITableViewCellAccessoryDetailDisclo sureButton` to the `accessoryType` property of the cell:

```
- (UITableViewCell *)tableView:(UITableView *)tableView
        cellForRowAtIndexPath:(NSIndexPath *)indexPath
    {
        static NSString *CellIdentifier = @"OMG_Its_a_Cell";

        UITableViewCell *cell = [tableView
            dequeueReusableCellWithIdentifier:CellIdentifier];
        if (cell == nil) {
            cell = [[[UITableViewCell alloc] initWithFrame:CGRectZero
                reuseIdentifier:CellIdentifier] autorelease];
            cell.accessoryType = UITableViewCellAccessoryDetailDisclosureButton;
        }

        cell.text = [nodes objectAtIndex:indexPath.row];
```

```
        return cell;
    }
```

Disclosure buttons handle user interaction separately from the cells in which they are embedded. To respond to user interaction for a disclosure button in an **accessory View**, define a `tableView:accessoryButtonTappedForRowWithIndexPath:` method in your table view's **delegate**:

```
- (void)tableView:(UITableView *)tableView
accessoryButtonTappedForRowWithIndexPath:(NSIndexPath *)indexPath
{
    // Forward the message to the tableView:didSelectRowAtIndexPath:
    // method. You can do anything you want here. You may want to
    // treat disclosure buttons separately from the cell, showing
    // a configuration screen instead of a detail screen, for example.
    [self tableView:tableView didSelectRowAtIndexPath:indexPath];
}
```

Disclosure buttons can also be used outside of tables. The simple clarity and ubiquity of disclosure buttons across applications enable users to understand that more information is available for the selected context:

```
- (id)initWithFrame:(CGRect)frame
    {
        if (self = [super initWithFrame:frame]) {
            self.backgroundColor = [UIColor clearColor];

            // Add a disclosure detail button to the view.
            disclosure = [UIButton buttonWithType:UIButtonTypeDetailDisclosure];
            [self addSubview:disclosure];
            disclosure.center = CGPointMake(270.0, 20.0);
            [disclosure addTarget:self action:@selector(disclose:)
                forControlEvents:UIControlEventTouchUpInside];


            // Other setup here.
        }
        return self;
    }
```

You can use custom buttons or other controls as accessories by adding them to the **accessoryView**. If you create a custom control for your **accessoryView**, you will need to use the target-action mechanism to capture the events generated by user interaction.

CHAPTER 8

Progressive Enhancement

Mobile devices are used in highly variable situations. Users are often outside the controlled, predictable confines of home or office, and they often have spotty or no access to networks. The basic operation of the iPhone doesn't require any network access (save for the telephony functionality), and custom applications should attempt to provide value in the absence of connectivity.

The term for developing baseline functionality with few external requirements and enabling additional functionality in more supportive contexts is *progressive enhancement*. The key topics or functional areas for using progressive enhancement on the iPhone are:

- Network connectivity
- Location awareness
- Accelerometer support
- Rotation support
- Audio support
- Vibration support

You will notice that some topics are related to communication while others focus on interaction. The general rule for connectivity is to require the bare minimum and provide additional features as conditions allow. For interactivity, it's best to assume the most chaotic environment for use and provide additional options for users who can interact with multiple fingers or both hands, or with the stability to use the accelerometer for physical gestures.

Use cases can help find functionality that you can shift from requirements to enhancements. You should define several use cases, with special consideration of the environments in which the user will interact with the application and with the device. Will the user be walking? Standing on a crowded commuter train? Flying on a commercial airplane? Will the application be useful on cold, blustery days, or while mixing drinks for friends?

Once you've created a few scenarios and identified the most likely, you will begin to recognize essential versus supplemental features. Including supplemental features gracefully is one of the more interesting challenges of UX programming.

Network Connectivity

The iPhone supports several forms of networking: Wi-Fi, 3G, and 2G EDGE. The iPhone will detect and use the fastest connection automatically. That is, if Wi-Fi is available, the iPhone will use it. If not, it will attempt to use a 3G connection. Failing that, it will use 2G.

In many circumstances, there will be no connectivity. For example, there is rarely a connection in the New York City subway system. Applications that rely on networking can often take steps to provide utility even in the absence of network connections. A great first step is saving any data currently in use when an application terminates. This might be web-based data like HTML pages, RSS feeds, or cached image files. Another good idea is handling any pending transactions between the device and remote servers. Network programming deals with data transfer and communication between machines. If a user loses network access while using an application, the application should act appropriately. In some cases, that will mean showing an alert that notifies the user of catastrophic failure. In many cases, though, a better option is simply to save any uncommitted transactions and present some sort of peripheral indication like an icon.

Figure 8-1 shows Shovel, an iPhone app for reading content from Digg. The screen displays the application as users see it when they have no network connection. A great enhancement to applications that deal with massive datasets might be to cache any data that loads, so users can revisit content in the absence of connectivity.

Maintain State and Persist Data

Applications often use network connectivity to transfer data from remote servers or systems. A good example of a networked application is an email program such as the Mail application. Email messages are transferred across the network to the iPhone and stored locally. Maintaining state lets users read their email even in the absence of a stable network connection that would allow the delivery of new messages. Many networked applications can remain useful and relevant to users in the absence of connectivity by displaying stored versions of transferred data.

There are many ways to save application data on the iPhone. One option is to structure your application data using a standard Cocoa data structure like an `NSArray` or `NSDictionary`, archiving the entire structure to a binary file. A second option is to use *property list* serialization, saving your data in a structured XML document. Finally, the most robust method of data storage on the iPhone is to use SQLite—a database that is included with the iPhone OS—or Core Data, a powerful framework for data management.

Figure 8-1. Shovel, an iPhone app for reading Digg content

Cache User Input

Reading data from the Internet is a very common requirement for mobile applications. A related requirement is to write data to network services. The presence and reliability of network services adds an element of risk for users: if connectivity drops or is absent, the state of both remote and local data becomes ambiguous. A standard pattern for ensuring state across network connections is to rely on return codes from the remote service. Applications can use representational state transfer (REST) over HTTP to simplify their communication and leverage return codes to ensure that local state and remote state match up where appropriate.

The communication protocols are less important than the use of networking. You should treat network connectivity as an enhancement rather than a requirement if possible.

What does this mean for an application focused on networking? Consider two examples: the built-in Mail application and any of the excellent Twitter clients for Cocoa Touch. In both cases, the ability to communicate with remote services is integral to the application. An email client must at some point—and preferably all the time—connect to the Internet. The same is true of a Twitter client. In such cases, networking is more than a friendly enhancement.

The requirement for Internet connectivity might be described in one of two ways:

- Internet connectivity is required to run the application.
- Internet connectivity is required to sync the local state of the application with remote services.

The second description lets us create a friendlier user experience. The ideal networked iPhone application is a client of network services, but networking is only one facet.

The Mail application solves the issue of network requirements. Email messages are synchronized to the device and remain readable in the absence of an Internet connection. Messages or replies are composed on the device and automatically saved locally in the Outbox (ready for delivery) or in the Drafts folder (marked as incomplete, but still saved). When a network connection is available, messages in the Outbox are sent to the appropriate mail server. If no connection is available, Outbox messages remain on the device, ready to go.

The ubiquity of this pattern in desktop email clients has set user expectations, and Apple delivered (no pun intended) by sticking to standard behavior. The same pattern is useful with any networked application. Any application can use the Outbox metaphor to address the natural separation of functionality: the user composes content, and the user transfers content. The two functions aren't necessarily linked together, though the latter relies on the former.

If you choose to address network writes this way, you should pay special attention to the mechanisms that notify users when messages are transmitted—and when transmission fails. One area in which Apple fails to deliver a smooth user experience is in the notification that mail cannot be sent. Mail treats a lack of connectivity as a failure state, and a full-screen modal alert is used to notify users. This alert abruptly interrupts the task at hand, such as composing a second message. Further, the user is required to click a button to dismiss the alert. A navigation prompt would be a more subtle mechanism for showing low-priority alerts. This example shows a prompt that displays the message "Offline" to a user:

```
- (void)showOfflinePrompt
{
    self.navigationItem.prompt = @"Offline";
    [self performSelector:@selector(hidePrompt) withObject:self afterDelay:2.0];
}

- (void)hidePrompt
{
    self.navigationItem.prompt = nil;
}
```

Reflect Connectivity Appropriately

Progressive enhancement is the art of using available resources to provide the best possible experience for users, and elegantly adapting to changes in resource availability. The way applications communicate available functionality and the state of data is a topic worthy of attention. An application can use several mechanisms to alert users to events or changes in state. Developers should carefully choose alerts that reflect the importance of an event. Most changes probably don't require abrupt interruption and the full attention of a user.

You can show simple status updates through iconography if the icons are universally familiar. A good example is the network status display on the iPhone. The display alternates between several states:

- Wi-Fi signal strength indicator
- 3G service indicator
- 2G EDGE service indicator
- No service indicator

If you create a full-screen application that uses network connections in a significant way, the native iPhone iconography may be a good foundation for your indicators and visibility.

Load Data Lazily

Lazy loading is a great data management pattern for progressive enhancement. It refers to the art of retrieving only the data you need to display or search at a given moment, which helps reduce latency and blocking of the main thread. The use of lazy loading is particularly helpful when working with large datasets or when traversing complex collections of objects. Working with `UINavigationController` instances to drill through hierarchical data provides a good opportunity for lazy loading, especially when reading data from an SQLite database.

To better understand lazy loading as it relates to navigation-based applications, consider a dataset that acts as a directory of companies and their employees. Figure 8-2 shows a simple database schema with example rows. Imagine hundreds of companies listed alphabetically in a `UITableView`.

The normal pattern for displaying this information would be to provide users with three levels of views: all companies, all employees for a selected company, and all information for a selected employee.

Implementing lazy loading for the screen that displays all companies would include selecting the `name` property of each `Company` record from the database with a query such as `SELECT name FROM companies;` and storing the results in an `NSArray` that can be used to populate table cells. Selecting a table cell representing a `Company` will push a new `UIViewController` onto the navigation controller stack, and a query will retrieve information for that controller. In this example, a developer would load all employee names for the company. A query such as `SELECT name FROM employees WHERE company_id = 1001;` would retrieve only the information needed for display and no more.

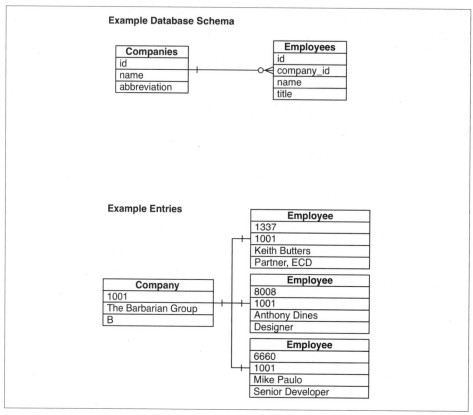

Figure 8-2. Sample database schema and example records

Developers accustomed to working on high-traffic database systems often have to balance the number of database calls with the *weight* of each call—that is, the amount of data being read and returned for a given query. Using SQLite on the iPhone is essentially immune to those concerns, and developers can safely retrieve additional data.

You can go further when loading data lazily by restricting not only the fields or associations loaded for an object, but also by loading only a subset of objects. The App Store loads search results in sets of 25 and displays a button at the bottom of the list that allows users to load an additional 25 records, adding the new objects to objects already in memory. This is different from pagination, which is a pattern for replacing subsets with new subsets. Modifications to the queries to leverage subset loading is simple:

```
SELECT name FROM companies ORDER BY abbreviation, id ASC LIMIT 25;

SELECT name FROM employees WHERE company_id = 1001 ORDER BY name ASC LIMIT 25;
```

Figure 8-3 shows the App Store offering to load yet another 25 tip calculators.

Figure 8-3. App Store search results with option to fetch additional records

It's possible to use lazy loading when working with file-based storage such as serialized objects or property lists, but doing so requires developers to manage multiple disparate files. For example, you might have all `Company` objects serialized in an `NSArray` object and saved to one binary file, and all `Employee` objects serialized in an `NSMutableArray` and stored in a separate file. This approach will work, but it's typically riskier than using SQLite because of the risk of file corruption, the inability to query for partial sets of records or partial attribute sets, and the lack of relational integrity that SQLite provides.

If your dataset is small—a few megabytes, for example—serializing a single collection is fairly safe. You won't have support for fast queries as with SQLite, but the overall management of your objects will be much simpler. Starting with the simplest option is a very smart move, but you should always keep an eye on the limits of the persistence mechanism you choose. If you find yourself splitting your object graph into multiple graphs and serializing them into separate files, it may be time to explore SQLite.

Lazy loading with Core Data

Core Data is a new feature for the iPhone with iPhone OS 3.0. With Core Data, developers access data using *managed objects*. A managed object is a special object in Cocoa that represents an entity in your application. When you work with Core Data, you can access objects directly. This approach differs from that of SQLite. With SQLite, you must query the database for data that is stored in rows of columns, like a spreadsheet.

Once retrieved, the rows must be converted to objects using tedious conversion processes, like assigning properties using the values in rows. With Core Data, you don't store objects directly as decomposed rows of data in a database. Instead, you insert an object or set of objects into a *managed object context*, which serves a similar purpose to a database. In fact, on iPhone OS, the underlying storage mechanism for managed objects is SQLite.

You access managed objects by asking the managed object context for a set of objects that fit certain criteria. Once fetched, objects can manage their own lazy loading using a mechanism known as *faulting*. For example, if you have a managed object representing a Company, and that Company has relationships representing a set of Employee-managed objects, you can fetch the Company from your managed object context first, then simply access the related Employee instances as properties of the Company. If the Employee objects aren't fully loaded into memory, Core Data will recognize the problem and fetch the missing data. This is a form of lazy loading that is built into Core Data.

The problem with lazy loading using Core Data is twofold. First, it's expensive. Your managed object context has to make a round trip to its underlying storage mechanism to read the data and assign it to managed objects. The process is similar to manual lazy loading using SQLite, though less work is required for application developers because Core Data supplies objects rather than raw SQLite records.

The second problem is that Core Data manages data locally and is highly optimized for the task, making the type of lazy loading required for SQLite unnecessary. Typical applications will have no need for lazy loading using Core Data, and if they do, the built-in faulting mechanism will load the required data as needed.

Core Data provides methods for fetching not only objects that meet certain criteria, but also related objects that are likely to be accessed and trigger lazy loading. Fetching sets of related objects, including all of their data, can consume available memory very quickly.

The power of Core Data comes at a cost when compared to a database or file-based serialization. Core Data is a robust framework. The balance between memory and processing time is a classic problem for UX programmers, but the flexibility of Core Data allows developers to design relatively efficient systems.

In short, Core Data deserves diligent study by anyone building Cocoa or Cocoa Touch applications.

Peer Connectivity with GameKit

The iPhone OS version 3.0 includes a framework called GameKit. Applications use the GameKit framework to enable peer-to-peer connectivity across Bluetooth. As the name implies, this feature is most useful when developing games, but non-game applications can also take advantage of GameKit to add fun features. For example, the App Store will soon include dozens of applications that let users exchange virtual business cards.

`GameKit` is an intriguing addition to the iPhone, and it will allow developers to create all kinds of interesting applications. Like all networking services, `GameKit` should be treated as an enhancement if possible. An application for exchanging business cards should provide functionality for users who are not in a peer-to-peer environment. A multiplayer game that uses `GameKit` sessions to foster group competition should provide a single-player mode for users unable to play against others.

Unlike simple Internet connectivity, peer-to-peer connectivity requires at least two users in close proximity with the same application installed and running. Additionally, those two users must agree to connect their devices via Bluetooth. Applications that treat peer-to-peer connectivity as an exciting option rather than a requirement will seem more valuable to users, leading to greater adoption.

Location Awareness

The iPhone gives developers several methods of detecting the physical location of a device. This is by far one of its most innovative features and has helped create many new business models and usage patterns. Developers have released location-aware applications ranging from social network clients to restaurant review tools to mapping applications to games.

The framework used for adding location awareness to an application is called Core Location, and it is a powerful and impressive toolset. It also comes with a price and some limitations. For example, finding the location of a device takes a lot of power, potentially leading to battery drainage. The more accurate the request, the more power is used. Battery life is a constant concern for developers of mobile applications, and Core Location forces developers to balance their needs with the desire to minimize power use. This is one reason that a turn-by-turn navigation system is a challenge for the iPhone.

Another trade-off when you add Core Location support to an application is the increased risk of feature fatigue. Feature fatigue is the feeling that a product or tool—such as an iPhone application—has too many options and is too complex. One of the differentiators of the Mac and iPhone user experience over other platforms is the clarity and sense of purpose in high-quality applications. Apple users tend to prefer smaller, more polished feature sets to larger, less elegant software.

A final downside to adding Core Location support is the result of a default security policy that requires users to grant permission to each application making location requests. This is something users become accustomed to over time, but for users new to the platform, it can be disconcerting. Figure 8-4 shows the permission prompt for the Maps application.

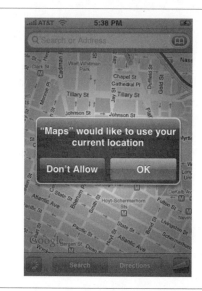

Figure 8-4. A Core Location permission prompt in the Maps application

Like many features of the iPhone, users can disable location support in the Settings application. The use cases you create for location-aware applications should factor in the possibility of users turning off support for Core Location. Marketing materials and App Store entries for applications that require Core Location should mention the requirement so that users who prefer not to share their location can make informed decisions. Figure 8-5 shows the option in the Settings application for disabling location support across all applications.

In cases where location services are disabled, users might be prompted to enable those services. Figure 8-6 shows the Maps application after launching on an iPhone with location services disabled.

None of these potential drawbacks is meant to discourage developers from adding location awareness to their applications. Rather, they're meant to be considered as part of due diligence for developers focused on creating quality user experiences. Just as some applications simply cannot function without a live network connection, location awareness is the crux of many applications.

Basic integration of location awareness is trivial. `CLLocationManager` is a class that resolves geographical location within a supplied radius. When a position is found, the class notifies a delegate object.

Core Location uses three methods of deriving location. The iPhone will choose the method that best suits the needs of the application. Developers can, in most cases, ignore the means used to locate a device and instead focus on the details of their request:

```
// Access the location manager
CLLocationManager *locationManager = [[CLLocationManager alloc] init];
```

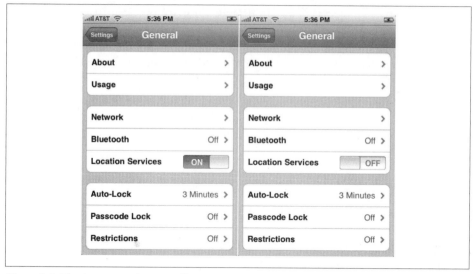

Figure 8-5. The Settings application allows users to disable location services for all applications

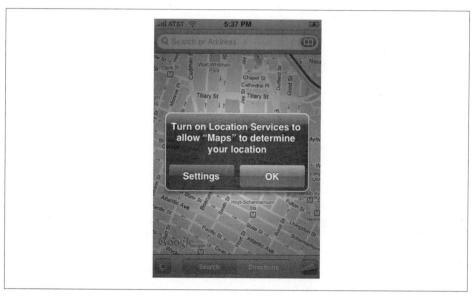

Figure 8-6. The Maps application with location services disabled

CLLocationManager lets you search for the current location. Each location manager should have a delegate assigned. A delegate can be any object that implements the CLLocationManagerDelegate protocol. The protocol is quite simple, consisting of only two event management methods: one representing a successful location search or refinement, and another representing a failure in deriving the location of a device.

A location manager object defines an instance property that can be used to set the desired accuracy for a Core Location request. The property is named, aptly enough, `desiredAccuracy`. The possible values for the desired accuracy are defined as constant double values of the type `CLLocationAccuracy`, and are described in the following list:

`kCLLocationAccuracyBest`

This value represents a request for the highest possible precision given the current conditions. This is a very expensive option because of the work required to derive the highest precision.

`kCLLocationAccuracyNearestTenMeters`

This value represents a request for accuracy within 10 meters.

`kCLLocationAccuracyHundredMeters`

This value represents a request for accuracy within 100 meters.

`kCLLocationAccuracyKilometer`

This value represents a request for accuracy within one kilometer.

`kCLLocationAccuracyThreeKilometers`

This value represents a request for accuracy within three kilometers. This option is preferable if you simply need to know the country, state, province, city, or town in which a device is located.

Setting the desired accuracy is simple:

```
locationManager.desiredAccuracy = kCLLocationAccuracyThreeKilometers;
// Set the delegate to handle the messages from the location manager
locationManager.delegate = self;
// Set the minimum distance the iPhone must move from its initial location
// before the delegate will be notified
locationManager.distanceFilter = 100.0f;
[locationManager startUpdatingLocation];
```

A location manager will continue to poll its sources in attempts to achieve an accuracy that is as close as possible to the stated `desiredAccuracy`. As it makes refinements, it will notify its delegate of changes, passing the new and prior locations to the delegate method `locationManager:didUpdateToLocation:fromLocation`. There is no guarantee that the desired accuracy will be achieved, or that it can be maintained in cases where the location manager is allowed to continuously seek new locations over time. When developing user interface elements that display the location, design for the unpredictable accuracy Core Location can provide. The Maps application handles the inherent imprecision of location services well. As the application refines location accuracy, the Maps application draws a circle with a radius that corresponds roughly to the accuracy of a given request. With each refinement, the circle animates and shrinks to show the new radius value. The radius represents the margin of error. Errors in precision still occur on occasion—showing that a user is in a lake instead of on a road circling the lake, for example. This imprecision does not hinder the user experience, however, because the large circles that refine over time communicate that accuracy is not guaranteed. Your design should take into account this variable precision, and should provide users with a clear understanding of the reliability of location data.

Setting the `distanceFilter` property tells the location manager to change the granularity of its polling and to only notify its delegate of changes in location when the device has traveled at least as far as the filter. This can help mitigate battery drainage, but the use of a distance filter isn't without power consumption.

In most cases, developers will need to retrieve location at least once and within certain accuracy tolerances. There is no property for the `CLLocationManager` class that can state an acceptable range of accuracy that, when reached, should stop polling. In addition, Core Location will cache the last known location of a device, and in most cases, the cached value will be the first reported location. This helps to improve the user experience by supplying an application with a relevant starting point for the update request. Each `CLLocation` instance has a `timestamp` property. It's important to check the `timestamp` for all location objects returned by a location manager, because they are often returned from cache or reported out of order as a side effect of the mechanism that resolves the location of a device. Location-aware applications that report an outdated location from cache will seem defective and frustrate users:

```
// Get the timestamp of a CLLocation instance

- (void)locationManager:(CLLocationManager *)manager
didUpdateToLocation:(CLLocation *)newLocation
fromLocation:(CLLocation *)oldLocation
{
    NSDate *locationTimeStamp = newLocation.timestamp;
    NSTimeInterval delta = [locationTimeStamp timeIntervalSinceNow];
    NSLog(@"The location timestamp interval was %d seconds.", delta);
}
```

Accelerometer Support

The accelerometer built into the iPhone has proven to be a very popular feature, especially for game developers. The accelerometer lets the entire device be used as an input device, with its orientation and movement in space transmitted to an application for custom handling. The ability to tilt and shake the device and control applications opens new possibilities for application developers. The user experience concerns for games are somewhat different from those for productivity, utility, or media-based applications. In most cases, accelerometer support for the latter types of applications is a novelty addition. If your application uses the accelerometer as the sole means of input for any piece of functionality, you may be alienating users who aren't in a position to—or who simply don't wish to—shake and twist their phone around to use a piece of software.

Accessing the accelerometer is a very simple process. The UIAccelerometer class represents access to acceleration events. You can grab an instance of UIAccelerometer with the singleton method sharedAccelerometer.

The instance requires a delegate assignment. The UIAccelerometerDelegate interface defines a single method that must be implemented to handle acceleration events. The method signature is accelerometer:didAccelerate:. Keep in mind that the UIAccelerometer construction method uses a singleton instance, which means there is always only one instance of the accelerometer in memory. As a side effect, only one delegate can be set for the accelerometer at a particular moment. This won't be a problem in most iPhone applications, but developers are capable of devising very interesting designs that might lead to unexpected results. For example, if you have two objects that set themselves as the delegate of the sharedAccelerometer, the first assignment will be overwritten by the second:

```
- (void)setupAccelerometer
{
    UIAccelerometer *accelerometer = [UIAccelerometer sharedAccelerometer];
    accelerometer.delegate = self;
    float updateInterval = 1.0f/24.0f;
    accelerometer.updateInterval = updateInterval;
}

- (void)accelerometer:(UIAccelerometer *)accelerometer
    didAccelerate:(UIAcceleration *)acceleration
{
    NSLog(@"The acceleration values (x, y, z) are: (%d, %d, %d)",
    acceleration.x, acceleration.y, acceleration.z);
}
```

Using the accelerometer outside immersive game environments will almost always introduce a new interaction pattern for mobile users. One such interaction pattern is shaking the device to clear the screen of user input. A simple way to shift such a feature from a requirement to a fun enhancement is to add a button that clears the screen in addition to using the accelerometer.

 The accelerometer in the iPhone and iPod Touch can be less accurate and more inconsistent than you may anticipate. The variability seems to differ among devices. There are two strategies to compensate for the lack of consistent accuracy. The first is to use the accelerometer as an enhancement rather than the sole input mechanism. Support touch-based options for playing games, adjusting orientation, or performing other orientation-based tasks. The second strategy is to test accelerometer code on multiple devices. The accelerometer is a fun and useful—if imperfect—feature.

Rotation Support

You can switch an iPhone between portrait and landscape orientations by turning the device in increments of 90 degrees. View controllers let developers handle changes in orientation as they occur, without requiring explicit access to the accelerometer. The most common behavior when orientation changes from the primary to the secondary mode is to redraw and refresh the layout of all onscreen views. This may simply involve changing the position and size of subviews. A change in orientation can also signal that interaction modes should be switched, providing a different view of the application altogether. For example, an application may present a table of football scores when the device is in portrait mode, and a chart of those scores when the device is rotated. The decision to change interaction patterns based on device orientation can be complicated. If the alternate view of the data is a supplement to the core functionality of the application, it can be an interesting example of progressive enhancement.

The decision to support rotation should take into account the ways in which users will interact with an application.

Views that display lots of text should consider supporting rotation, because iPhone users are accustomed to rotating their devices to increase either text size or the character count for each line of text. Both scenarios can improve clarity when reading dense text on a small screen.

Supporting basic view rotation is very easy. The `UIViewController` class defines a suite of methods and properties for handling orientation.

The `interfaceOrientation` property of each `UIViewController` instance represents the current interface rotation for the view. You should override the `shouldAutorotate ToInterfaceOrientation:interfaceOrientation` method and return `YES` for all conditions in which rotation should occur:

```
- (BOOL)shouldAutorotateToInterfaceOrientation:
(UIInterfaceOrientation)interfaceOrientation
{
    return (interfaceOrientation == UIInterfaceOrientationPortrait ||
    interfaceOrientation == UIInterfaceOrientationLandscapeLeft);
}

- (void)willRotateToInterfaceOrientation:
(UIInterfaceOrientation)toInterfaceOrientation
duration:(NSTimeInterval)duration
{
    // Stop any animations, hot UI objects, or redraw operations
    // Prepare for rotation
}

- (void)didRotateFromInterfaceOrientation:
(UIInterfaceOrientation)fromInterfaceOrientation
{
    // Restart any animations, hot UI objects, or redraw operations
}
```

The autorotation mechanism is based on four distinct points. Developers can take advantage of any combination of those points to perform additional functionality. For example, you may want to swap a particular graphic from the screen midway through a rotation sequence. You can override the following four methods in your `UIViewController` subclass to support custom rotation logic:

`willRotateToInterfaceOrientation:toInterfaceOrientation:duration`
> Called before the autorotation sequence begins. Use this callback to prepare for rotation by pausing expensive redraw operations, disabling touch-sensitive controls or views, and, if necessary, swapping out the main view with another to be shown during the rotation animation.

`willAnimateFirstHalfOfRotationToInterfaceOrientation:duration:`
> Called before the first half of the rotation—that is, the exit rotation for the current layout—is animated. Any header and footer bars at the top and bottom of the view animate out immediately after this callback.

`willAnimateSecondHalfOfRotationToInterfaceOrientation:duration:`
> Called after the first half but before the second half of the rotation animates. Any header and footer bars animate back into the frame immediately after this callback is triggered.

`didRotateFromInterfaceOrientation:fromInterfaceOrientation`
> Called after the autorotation sequence completes and all views are in place. Use this callback to enable any paused effects, touch-sensitive controls, and status indicators.

The iPhone device itself often disappears into the background of a user's attention, allowing the current application to become the entire experience. Consider the minor disruption of that focus when requiring users to rotate the device. The act of pulling back from the device, rotating it, and resetting focus may seem like a minor distraction, but it's significant enough to merit consideration. This concern becomes more relevant in cases where rotation is required, rather than simply supported as a progressive enhancement feature.

Audio Support

Many applications provide support for ambient or event-based audio in the form of background music or sound effects. The use of sound is particularly strong in games— both immersive 3D experiences and tile- or sprite-based 2D games. Audio support makes a lot of sense for Cocoa Touch applications. After all, the iPhone is more a descendant of the iPod than a cellular phone, and the iPod Touch is positioned primarily as a media player.

Users often wear headphones when working with the devices, making the barrier to experience quite low. The media player role of the devices can dissuade developers from adding audio support to their applications—after all, when users are wearing headphones, they are likely listening to music or making a phone call and are thus opening an application for some secondary purpose.

There are two approaches to handling audio with Cocoa Touch. The first approach is to interrupt audio playback by other applications. For example, if you develop an application that records sound using the microphone, you may want to stop playback of music running in the iPod application.

The second approach is to mix audio from your application with audio from other running applications, such as the built-in iPod application. This approach treats audio more as an enhancement than a requirement, and is preferable for applications that don't focus primarily on audio. Using the sound recorder example, you might allow the iPod to continue to play music so that users can record themselves singing along to a song.

You will notice a recurring theme in deciding how to handle various user experience concerns. User expectation should be a particularly significant consideration. Important questions to ask yourself are:

- How should my application behave if the phone rings?
- How should my application treat music that is playing when the application is launched?
- How should my application treat phone calls occurring when the application is launched?
- What is the best default behavior where sound effects are concerned?
- What controls should play sound?
- Should the application respect global settings regarding sound effects?
- Does the core function of the application depend on users hearing all audio?
- Are only some sounds essential for the proper use of the application?

The simplest way to play short audio on the iPhone is to use System Sound Services. All sounds played through System Sound Services will mix with audio from other applications and will obey the standard behavior around the hardware Ring/Silent switch and volume rocker. System Sound Services should be used for playing non-essential audio effects, such as startup or alert sounds.

There are two functions for playing sounds through System Sound Services. They are almost identical in their use and operation. The difference is that one function will trigger a vibration in lieu of playing audio if the phone is silenced, whereas the other will not. When choosing these functions, developers should take care to examine user expectations:

`void AudioServicesPlaySystemSound (SystemSoundID inSystemSoundID)`
> Play a sound resource as identified by a `SystemSoundID`. If the Ring/Silent switch is set to silence the ringer, the sound will not play. This option is preferable for all

short, non-essential sounds, such as sound effects for controls or game feedback sounds.

void AudioServicesPlayAlertSound (SystemSoundID inSystemSoundID)

Play a sound resource as identified by a SystemSoundID. If the Ring/Silent switch is set to silence the ringer, the phone will briefly vibrate. This function should be used only when it's vital to communicate to the user that an event has occurred, because vibration is a disruptive form of feedback that can't be ignored. If a user has silenced their phone, they likely want to avoid interruptions. Interruptive errors are a good use of this function.

The following example plays an audio response to a network operation. If there is an error, an alert sound—or vibration, if the device is silenced—is played. In a real application, you would probably cache the SystemSoundID variables in your instance:

```
// Assumes a declaration of instance variables:
// SystemSoundID successSound, errorSound
- (void)performAudioNotificationForStatus:(BOOL)success
{
    NSString *successPath = [[NSBundle mainBundle]
        pathForResource:@"success" ofType:@"caf"];
    NSString *errorPath = [[NSBundle mainBundle]pathForResource:@"failure"
        ofType:@"caf"];
    OSStatus error;
    if(success){
        if(successSound == nil){
            error = AudioServicesCreateSystemSoundID(
                (CFURLRef)[NSURL fileURLWithPath:errorPath],
                &successSound
            );
        }
        AudioServicesPlaySystemSound(successSound);
    }else{
        if(errorSound == nil){
            error = AudioServicesCreateSystemSoundID(
                (CFURLRef)[NSURL fileURLWithPath:successPath],
                &errorSound
            );
        }
        AudioServicesPlayAlertSound(errorSound);
    }
}

- (void)dealloc
{
    AudioServicesDisposeSystemSoundID(successSound);
    AudioServicesDisposeSystemSoundID(errorSound);
    [super dealloc];
}
```

If an application needs to play sounds longer than 30 seconds or play non-trivial audio, you can take advantage of several more advanced audio playback frameworks and tools:

OpenAL
> Used especially for spatial sound programming. OpenAL is popular in game programming or in applications used for playing audio with spatial variation. The OpenAL FAQ for iPhone OS, bundled with the iPhone SDK documentation, is a good starting point for learning about OpenAL.

Audio Queue Services
> Used as a straightforward way to record or play audio. The Audio Queue Services Programming Guide, bundled with the iPhone SDK documentation, is an excellent resource for learning more about Audio Queue Services.

`AVAudioPlayer` *class*
> The `AVAudioPlayer` class, bundled with a delegate protocol, was added to iPhone OS 2.2 as part of the `AVFoundation` framework. The `AVAudioPlayer` class provides robust functionality around playing audio files. Developers can use the class to play multiple sounds, play sounds with distinct levels, loop sounds, seek through files during playback, and gather data that can be used in visualizing the power levels of audio channels. The `AVAudioPlayerDelegate` protocol defines methods used for handling interruptions to playback, such as from an incoming phone call, along with expected error, start, and stop events.

Audio units
> Audio units are audio processing plug-ins. You can read about audio units in the Audio Unit Component Services Reference in the iPhone SDK documentation.

When using any of these more complex audio programming resources, you can control the way your sounds mix with other applications using audio sessions. The Audio Session Services Reference documentation, included with the iPhone SDK, describes the use of audio sessions. Audio sessions are used to define the conditions of operation for audio handling on the iPhone OS. Each application is given an audio session object that acts as an intermediary between the application and the operating system. Developers set properties of the audio session for their application to define how the iPhone OS should make decisions about audio playback in response to the myriad conditions that affect the mobile user experience. Typical events with consequences for audio playback are:

- Incoming phone calls
- Terminating phone calls
- Playing music with the iPod application
- User interaction with the Ring/Silence switch
- Users switching between the headphone jack output and the built-in speaker
- Users interacting with the Sleep/Wake button
- The auto-sleep timer expires and the device goes to sleep

There are lots of ways to customize audio sessions, and Apple has provided several preset collections of settings—called *categories*—that solve the most common use cases. Though only one category can be assigned to an audio session at any given time, the category is mutable and can be changed for different circumstances. The audio session categories and the corresponding UInt32 constant used to set the category are:

UserInterfaceSoundEffects
> Used for applications that provide simple, short, user-initiated sound effects, such as keyboard taps or startup sounds. Sounds will mix with audio from other applications and will obey the Ring/Silent switch. Identified by kAudioSession Category_UserInterfaceSoundEffects.

AmbientSound
> Used for slightly longer sounds that can safely obey the Ring/Silent switch without hindering the core functionality of the application. Sounds will mix with audio from other applications and will obey the Ring/Silent switch. Identified by kAudio SessionCategory_AmbientSound.

SoloAmbientSound
> Used similarly to the AmbientSound category, except that the audio does not mix with audio from other applications. Identified by kAudioSessionCategory_SoloAm bientSound.

MediaPlayback
> Used for audio or audio/video playback. All other applications are silenced. Media Playback sounds do not obey the Ring/Silent switch. Identified by kAudioSession Category_MediaPlayback.

LiveAudio
> Used for applications that simulate music instruments, audio generators, or signal processors. All other applications are silenced. LiveAudio sounds do not obey the Ring/Silent switch. Identified by kAudioSessionCategory_LiveAudio.

RecordAudio
> Used for applications that record audio. All other applications are silenced. Record Audio sounds do not obey the Ring/Silent switch. Identified by kAudioSession Category_RecordAudio.

PlayAndRecord
> Used for applications that both play and record audio. All other applications are silenced. PlayAndRecord sounds do not obey the Ring/Silent switch. Identified by kAudioSessionCategory_PlayAndRecord.

Setting a category isn't difficult. This example sets a session to the AmbientSound category:

```
- (void)prepareAmbientAudio
{
    OSStatus result = AudioSessionInitialize(NULL, NULL, NULL, self);
    if(result){
        NSLog(@"Ambient audio could not be prepared. Consider this suspicious.");
```

```
    }else {
        UInt32 category = kAudioSessionCategory_AmbientSound;
        result = AudioSessionSetProperty(
            kAudioSessionProperty_AudioCategory,
            sizeof(category),
            &category
        );
        if(result){
            NSLog(@"Could not set audio session category. This is no good. Fix it!");
        }else {
            result = AudioSessionSetActive(true);
            if(result){
                NSLog(@"Could not set audio session to active status. Fix it!");
            }
        }
    }
}
```

Audio programming possibilities on the iPhone are shockingly robust for a mobile device. Of course, with great power comes great responsibility—so be considerate of user expectations when adding audio to applications, but don't hesitate to create innovative applications. Let users decide how and when audio should contribute to their experience.

 Like sound, vibration can be either an excellent enhancement or an annoying deal breaker for users of an application. The mobile HIG describes user expectations as they relate to vibration, and Apple advises developers to use vibration only when appropriate. You can trigger vibration by calling the AudioServicesPlaySystemSound function and passing a constant SystemSoundID called kSystemSoundID_Vibrate:

```
- (void)vibrate
{
    AudioServicesPlaySystemSound(kSystemSoundID_Vibrate);
}
```

CHAPTER 9

UX Anti-Patterns

A design pattern is a common approach to solving a problem. All developers apply design patterns to their work, even if they don't realize it. There are countless books that cover design patterns for software engineers and designers, and this book is no exception. An anti-pattern is a common mistake when addressing a problem. Anti-patterns are the flip side to design patterns; they are basically design patterns that should be avoided because they cause at least as many problems as they solve.

Most design patterns have names; you've likely encountered the Delegate, Chain of Command, Singleton, Observer, and Model-View-Controller patterns. Anti-patterns are often named as well. This chapter is an exploration of anti-patterns found in many applications in the App Store.

It's important to keep in mind, however, that an anti-pattern doesn't equal a bad application. As you'll see in this collection of common anti-patterns, many excellent applications include a problematic feature or two. Applications, like standards, evolve over time. The most important thing a developer can do is to constantly consider the perspective of the user and to recognize patterns—both good and bad.

Billboards

Chapter 5 covered launch screens and the *Default.png* pattern built into the loading of Cocoa Touch applications. Despite the Human Interface Guidelines' recommendation that launch screens act as visual placeholders for loading applications, a large number of applications use the delay as an opportunity to perform a bit of branding. In the cooperative interaction model, applications should feel as though they aren't opening and closing, but rather pausing and unpausing. Users should feel that they are cycling through a suite of highly responsive applications, rather than having their experience interrupted by a full-screen logo, advertisement, or "About" screen.

For example, Tweetie is an application that reveals the user interface progressively, as described in the HIG. Figure 9-1 shows the opening sequence for Tweetie.

Figure 9-1. The opening sequence for Tweetie

Imagine a user who launches her email client and reads a message with a link to a website. Upon clicking the link, Mail closes and Safari launches with the site URL. The user decides to email the link to another person, invoking the "Mail Link to this Page" command in Safari. Safari quits and the Mail application launches with a new message pre-populated with the link to the site.

The flow between the two applications is highly cooperative. Each performs a specific role and works with the other to let the user accomplish her goal of sharing an interesting link. The user sees that the two applications are passing focus back and forth.

Now imagine the same flow if each application were to show a splash screen with a logo or advertisement. The user experience would suffer because the illusion of cooperation and seamless flow between applications would be destroyed.

Figure 9-2 shows the launch and main screens for the Streaks application. Though it is beautiful and informative, the launch screen can interrupt the flow between applications if users switch between Streaks and other applications many times in a single session.

The Wurdle application takes a similar approach to the loading screen. Figure 9-3 shows the launch sequence for Wurdle.

Immersive games are somewhat of an exception, but even gamers rarely wish to watch the same branded splash video each time they launch a game. The focus should always be on users, and all possible efforts should be made to improve their experience. Some applications abuse the `sleep()` function to purposely display the launch screen longer than the application requires.

Figure 9-2. The launch and main screens for Streaks

Figure 9-3. The launch sequence for Wurdle

Some applications look very different at each launch. An application that uses a tab bar may be programmed to store the number of the tab that is displayed when the application is closed, allowing the application to programmatically restart with the same tab focused. If each tab looks very different, it may be difficult to design a launch screen that smoothly follows the pattern Apple built into the launch sequence. Developers can often reduce the graphics to a bare minimum for the launch screen, even using a solid color or image as the background for all tabbed views.

Some applications use the concept of themes to let users change the colors and layout of elements on screen. For example, Twitterrific lets users choose from several very different themes. There is no common color or background among views, and even the system status bar differs among themes. In such extreme cases, a simple splash screen that transitions to the custom color scheme using animation could be an option.

Misused animations can be as disruptive as billboards, so any such transitions should be subtle.

The goal of a launch screen is to create the illusion that an application loads instantly. Billboards are counterproductive to that goal. If the only satisfactory option is a distinct launch screen, developers should pass on the opportunity to create a loud, branded billboard and opt for something users won't mind seeing multiple times per day.

Sleight of Hand

The layout of Cocoa Touch user interfaces can have a big impact on usability. This is especially true in regards to user interface controls. A control is any interface element that causes the application to do something when a user touches the screen. Standard controls include buttons, sliders, and switches. Table cells can act as controls, as can entire views.

The Multi-Touch interface on the iPhone and iPod Touch requires that applications implement virtual controls. Unlike physical knobs or buttons, virtual controls provide no tactile (or *haptic*) feedback. A great example of tactile feedback on a hardware control is the raised bumps on the F and J keys of a standard English QWERTY keyboard. These bumps indicate to users that their fingers are in the proper place on the keyboard. On a touchscreen, muscle memory is built with repeated use, but orientation is based purely on the portion of the screen a user is touching.

The sleight of hand anti-pattern occurs when developers change the perceived meaning of a touch-sized area of the screen by placing controls that perform different actions in the same place across views.

Apple provides a good example of the sleight of hand anti-pattern. The iPhone displays one of two screens in response to an incoming call. In both scenarios, the screen shows the name of the caller and, if available, a photograph assigned to the incoming phone number. If the iPhone is locked, users see a large slider labeled "slide to answer" and a slider handle that moves from left to right. Sliding will unlock the phone and answer the incoming call. Figure 9-4 shows the incoming call screen on a locked iPhone.

If the phone is unlocked, the screen shows the same standard incoming call screen but replaces the large slider with two buttons: one to decline the call and one to answer the call. The Decline button is on the left, and the Answer button is on the right.

In this example, the sleight of hand occurs when a user who often keeps his phone in a locked state receives a call when the phone is in the unlocked state. The user is probably trained to touch the left side of the screen and quickly swipe to the right to answer a call. This causes problems when the phone is in an unlocked state because the user will instinctively touch the lower left part of the screen, inadvertently hitting the Decline button.

Figure 9-4. An incoming phone call on a locked iPhone

The sleight of hand anti-pattern commonly appears in navigation controllers. When working with a navigation controller, a user will often tap the upper left area of the screen multiple times in quick succession to move up a hierarchy of information. Users learn to perform this type of operation because so many applications use Apple's navigation controller design pattern to display hierarchies.

Navigation-based applications use a navigation bar at the top of the screen. The navigation bar usually has a button on the left that takes a user up one level, assuming she isn't already on the topmost or root level of information. If a user is on the root level, the left button area of the navigation bar should probably be empty to keep users who repeatedly tap a back button from accidentally performing an action due to a control taking the place of the button they are tapping.

Figure 9-5 shows the sleight of hand anti-pattern in the Facebook application. Tapping the "Photos of Toby" item shows the "Photos of Toby" second screen. The "Toby" back button on the second screen transitions the user back to the first screen and swaps the back button for a Logout button. An extra tap on the top-left portion of the screen can trigger the logout process. The Facebook application handles this well by requiring a user to confirm that they wish to log out, but the issue could be avoided completely by moving the Logout button to the right side of the navigation bar.

This can be a tough pattern to avoid because there may be perfectly valid reasons for a control taking a particular spot on a particular screen. Designers can find balance by looking at how users interact with an application holistically instead of designing layouts as a set of decoupled screens.

Figure 9-5. A sleight of hand in the Facebook application

Bullhorns

`UIKit` offers a lot of flexibility in the way an application notifies users of events, errors, or state changes such as losing Internet connectivity. The most obvious is a class called `UIAlertView`, which displays a pop-up box that shows the alert message along with one or more buttons that must be clicked to dismiss the box. Alerts are modal, which means they are presented on top of the current view, often blocking all interaction with the current application and stealing the focus of a user.

Another type of modal alert is an action sheet, invoked with the `UIActionSheet` class. An action sheet is similar to an alert box in its effect, but it has a different look. Action sheets can contain many buttons and shouldn't display text. Designers use action sheets primarily to let users confirm an action. For example, the Mail application uses action sheets to confirm the deletion of an email.

The bullhorn anti-pattern is the overuse of alerts that require users to make choices or otherwise manually dismiss the alert. In other words, the pattern causes disruptive notices that shouldn't merit such brute-force methods.

Examples of user interaction that probably don't require taking control of the interface and monopolizing user attention are:

- Switching from Wi-Fi to 3G or 2G service, or losing connectivity.
- Touching an illegal square in a chess game.
- The end of a turn or round in a game.
- A form submission failing due to a typo.
- A successful post to a web service.

- Any successful operation.

Events do occur that may have significance for users, but you should still avoid shouting the news. The cell signal strength indicator or Internet access indicator in the status bar are excellent examples of passive indication. Other good options are as follows:

- Inline messaging and highlighted form fields can signal problems with user input.
- Subtle sounds can communicate that a game piece has been moved to the wrong space. Even better, you can simply reject the move and animate the piece back to its original location.
- You can use a passive visual cue like a checkmark instead of a progress indicator when a Twitter submission succeeds.
- You can change the text of a button if a download fails. For example, change "Update Feeds" to "Retry Feed Update" if the first attempt fails.

Figure 9-6 shows a timeout error in Tweetie, an excellent Twitter client. An alternative to using an alert could be to display a concise but meaningful message in the main view, perhaps with an option to try again.

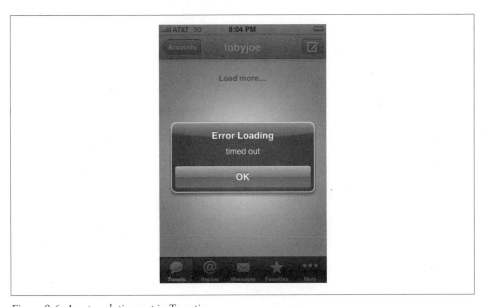

Figure 9-6. A network timeout in Tweetie

Figure 9-7 shows two alerts used in the Black and White application.

Generally, alerts should be no more forceful than the impact of the event. An alert box not only sends a device-level notice to users, but it also requires users to acknowledge the error by clicking an "OK" button. There are excellent cases for such a disruptive notice, but such cases are rare.

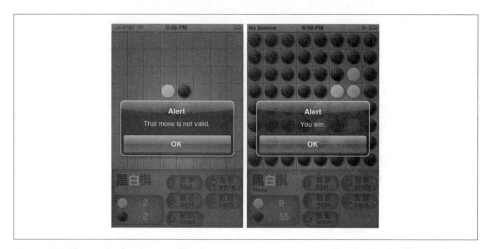

Figure 9-7. The result of making an illegal move or winning a game in the Black and White application

The iPhone elegantly communicates the current battery level to users. The battery level indicator is a small, passive icon in the status bar. When the available power drops below 20%, an alert box notifies the user. This is especially useful when a user is immersed in a full-screen experience, such as an OpenGL game or a video podcast. A second alert shows when the battery reaches 10%. Apple chooses not to bother users unless there is a risk of the battery dying. The same judicious restraint should be used for alerts in iPhone applications.

Figure 9-8 shows subtle messaging of network failures in two Apple applications: Stocks and App Store.

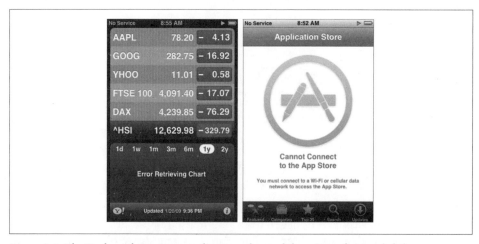

Figure 9-8. The Stocks and App Store applications show subtle notices of network failures

App As OS

Apple requires that applications not compete with the suite of applications that make up the default user experience. Applications are often rejected for duplicating functionality already supplied by Apple. This means that the market will likely never see real competition for the iPod application, Mail, or Safari, among others.

There are many ways of looking at this policy. One non-cynical view is that competing applications hurt the overall user experience. Instead of building a competitor to Safari, for example, developers can provide links that will open Safari, letting the most capable application handle the task at hand. As another example, you can enable users to send an email by making calls to load a URL with the `mailto://` URL scheme, which is intercepted by the Mail application. With the right arguments in the URL, the Mail application will instantly craft a message with a subject, a recipient address, and even a message body.

iPhone OS 3.0 lets developers go a step further than a `mailto://` link and incorporate a standard screen for composing and sending email messages. Apple seems to be selectively moving functionality from applications to frameworks, giving developers both options for performing common tasks while ensuring a consistent user experience for those tasks.

Despite Apple's policies, some applications with system-provided functionality make it into the store. It seems that Apple approves a few uses of duplication—all of them limited and minimal. The most common features that sneak past the censors are bundled browser functionality using `WebKit` and built-in SMTP support for sending emails directly from an application designed for iPhone OS prior to version 3.0.

The problem with the limited subsets is that users grow accustomed to certain tools, options, and features. When viewing a web page, for example, it's very common to use the full Safari feature set. Users pinch to scale, double-tap DOM elements to zoom in and out, hold links to see the target URL in a callout bubble, rotate the device for easier reading, bookmark pages, and email links to people. Most applications that present a `WebView` to load pages offer very few of these operations. The controls also differ with each implementation, leading to confusion and a learning curve for a function that users see as primary to the iPhone: browsing the Web. It's certainly ironic that Apple allows unpleasant implementations of duplicate features—especially given the justification that it's all in service to a better user experience.

Figure 9-9 shows the Shovel application, a client for Digg. The developers present users with the option to open the given news story in Safari, which is an excellent use of cooperative design.

Figure 9-9. Cooperative design in the Shovel application

Interestingly, if a user taps the headline of the story instead of using the action icon, Shovel opens the web page in an internal `WebKit` browser. As with other built-in browsers, the more compelling and user-friendly features of Safari are omitted. Figure 9-10 shows internal browsers used in Shovel, LinkedIn, and TED.

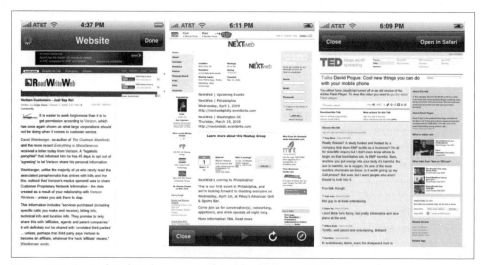

Figure 9-10. Internal WebKit browsers in Shovel, LinkedIn, and TED

Internal browsers can be useful, but they can cause confusion and frustration if familiar features are missing. If you include a browser screen in your application, you should provide at least the most essential functionality: next and back buttons, a reload button, and an option to open the current page in Safari. Give your users control over the use of limited embedded browsers through application settings.

Perhaps if Apple allowed developers to create robust alternatives to Apple applications like mobile Safari, the pattern would prove useful instead of being a frustrating anti-pattern. For now, however, the best practice is to implement cooperative functionality that lets users accomplish tasks using the tools they already know, and to focus custom development efforts on innovative problem solving, gaming, or media display.

Spin Zone

At some point in the development lifecycle of an iPhone application, a developer must decide how to handle support for rotation of the device. The basics of rotation support are very simple, though each view in each application has different, often complicated, considerations. How will interface elements scale? How will typography react? How should content react when the rectangle defining its boundaries changes dimensions and orientation?

Some developers choose to ignore rotation events, keeping the orientation of their application pinned to the screen and non-reactive to changes in the device. They always have the option to enhance the application at a later date, should users request rotation support. Other developers implement full rotation support, with rich animation and intuitive layout responses that keep their application usable from any angle. Both of these options contribute to a reliable user experience.

The spin zone anti-pattern occurs when an application implements rotation support in an ad-hoc fashion. For example, perhaps the application supports rotation only in one direction—90 degrees clockwise but not counterclockwise. Another example of the anti-pattern is arbitrarily—from the point of view of the user—supporting rotation for different screens. Worse still is requiring rotation for some screens, but not others.

The reasons that partial rotation support is damaging to the user experience aren't mysterious. If a user has to constantly check whether each screen supports rotation, it brings the device to the forefront and makes the hardware the center of attention, taking users out of the otherwise immersive experience. Users should not be forced to use a "try and see" approach to constantly test support for capabilities in an application.

The Bouncer

The App Store contains many great applications that act as clients for web applications. For example, there are several excellent applications for interacting with social services like Twitter or publishing content to blogs.

A requirement for most Cocoa Touch web application clients is to provide authentication credentials, such as a username and password, for an account on the remote web service with which the mobile application interacts.

The bouncer anti-pattern occurs when the Cocoa Touch application requires authentication credentials for operation, but doesn't offer the option for creating those credentials inside the application. Instead, the application opens Safari and forces registration through a web page, or worse, offers no option for registration at all.

There are cases in which registration from a mobile application makes little sense. For example, an application that allows you to check the balance in your savings account or retirement fund would require that the accounts already exist at the financial institution. The WordPress application requires external authentication using a WordPress-powered blog hosted on the Web. Because of this externality, creating an account using only the iPhone application would be impossible. Rather than simply blocking access for new users, the WordPress application allows access to information about the application and about WordPress. Figure 9-11 shows the WordPress application.

Figure 9-11. The initial screen for new users of the WordPress application

Most web applications are technically capable of allowing registration from any HTTP compatible client, including Cocoa Touch applications. The ability to register from an application may be limited by the terms of service. If policy prevents registration, applications should state as much and offer details on acceptable registration processes or sites.

The LinkedIn and Facebook applications are good examples of applications that implement the bouncer anti-pattern. Figure 9-12 shows the LinkedIn and Facebook applications after launch. Each application assumes that the user is familiar with the purpose of the application and either has or knows how to attain credentials to use the application.

Figure 9-12. The welcome screens for the LinkedIn and Facebook applications

An excellent example of an application that avoids the bouncer pattern is Foursquare. The authors of Foursquare developed the application with usability in mind and avoided the bouncer pattern in a simple, clean manner. Figure 9-13 shows the welcome screen for the application.

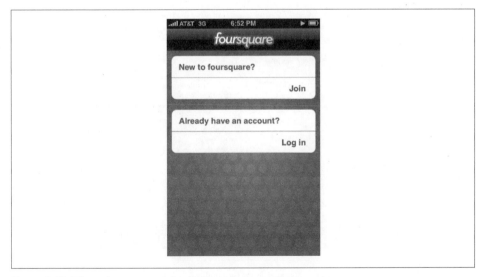

Figure 9-13. The welcome screen for the Foursquare application

Gesture Hijacking

The iPhone and iPod Touch train users to use specific gestures or finger motions to trigger certain actions. Swiping a finger from left to right across a cell in a table should tell the application to delete that cell and the information it represents. Pinching two fingers together while touching the screen should zoom the current view out, or reduce its size, proportional to the distance the fingertips move toward each other. A reverse-pinch should trigger the inverse effect.

Gesture hijacking occurs when a developer changes the meaning of these well-known gestures. Often, the implementation is innocent. A developer may simply want to introduce a novel method of interaction into an application and may not realize that they have collided with the user expectation of consistency. For example, an implementation may rework the built-in handling of swipes in table cells and, instead of triggering a deletion, allow users to "flick" the interface to the left to drill down to the next level in a hierarchy. This might be interesting, but could frustrate users accustomed to interacting with tables in the normal way.

Figure 9-14 shows an example of gesture hijacking in Tweetie. Swiping a cell in a table doesn't trigger the standard deletion behavior; instead, it uses nice animation effects to reveal an underlying set of controls relative to that table cell. The supplemental controls do not include a control for deleting the cell or its associated object—a fact that pushes the feature into the hijacking category. Though the feature is very interesting, it contributes to an uncertainty around an otherwise standard gesture. This may be great for the application, but can be detrimental to the overall user experience.

Figure 9-14. The gesture hijacking pattern as used in Tweetie

Users welcome interesting new implementations of Multi-Touch technology and understand that iPhone and iPod Touch applications are often quite innovative. Still, novelty should always be balanced with expectations. The swipe, flick, pinch, and reverse-pinch gestures should be used consistently, and applications should introduce new gestures for new features.

Memory Lapse

Network-aware applications, such as RSS readers or clients for sites like Twitter and Digg, inherently depend on an Internet connection. Without web access, the data they display is off-limits.

The first time users launch such an application, they probably expect a load operation of some sort, preferably with a progress indicator to show that the application is doing something in the background. Notice should be given not only that a download is in progress, but also whether or not the download is successful. In most cases, a successful operation will simply display the downloaded data. Failures can be communicated with an icon, as in Twinkle, or with passive messaging, as in Shovel.

Figure 9-15 shows the first launch and network sync for the TED application. The translucent overlay and status indicator tell the user that the application is actively communicating with the TED web services, while the underlying screen gives a preview of the type of content. This approach is preferable to that taken by many applications that fail to store information when applications close, requiring users to have a network connection to continue reading content they have already downloaded at least once.

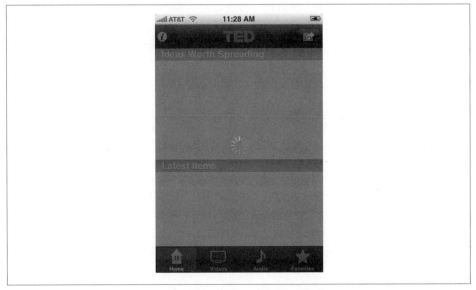

Figure 9-15. The first launch and network sync for the TED application

After the first successful download, user expectations change. Users expect that applications will store downloaded data between launches. This expectation comes from the overall feeling on the iPhone or iPod Touch that applications don't start and stop, but rather are paused, waiting for users to switch them back to the forefront.

Generally, an application should not forget data—especially the most recent data successfully downloaded. Users should not be penalized for changing albums, making a phone call, answering a text message, or otherwise closing an application. The application should provide the same functionality when reopened as when it was last closed, despite access to networks.

Figure 9-16 shows the result of launching both Shovel and Tweetie in the absence of a network connection. Users would likely benefit from an option to read content from the most recent successful sync.

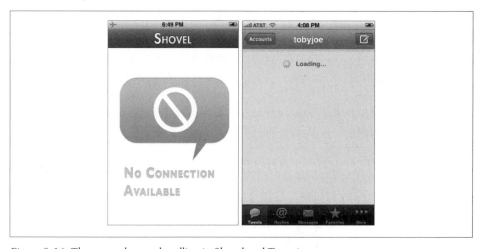

Figure 9-16. The network error handling in Shovel and Tweetie

The Facebook application uses a disk cache and does an excellent job of avoiding the memory lapse anti-pattern. Any data loaded into the Facebook application is saved to disk and redisplayed when the application is opened. This approach works very well to provide value to users without a network connection. It also has the side effect of improving speed for users who are online, because data is only pulled from the network if it is outdated or missing; otherwise, it's served from the cache.

When the Facebook application is asked to display content that is not cached and there is no network connection, users are given a clear but non-disruptive notice. Figure 9-17 shows the Facebook application responding to a request for data that cannot be located.

Figure 9-17. The error screen for unavailable data in the Facebook application

Developers of networked applications should work to avoid the memory lapse anti-pattern. Mobile applications are far more likely than desktop applications to encounter unfavorable network conditions. A good rule of thumb for data persistence is to ensure that your application can be opened on an airplane or on the New York City subway and still provide value to your users.

The High Bar

Progressive enhancement methods are ideal for providing supplemental functionality to users. Software developers can delight users by including supplemental features that add fun, novelty, or additional efficiency for task-oriented users.

The high bar anti-pattern is present in applications that limit the ability to accomplish tasks for users not meeting certain external criteria. Examples are applications that:

- Force users to shake their device to delete information.
- Require network connectivity for operations that could be accomplished offline.
- Require Multi-Touch input for essential operations and fail to provide secondary controls for users who are disabled or have only one free hand.
- Require precision input by presenting small input controls or small margins of error for touches and gestures.

The opportunities for the high bar anti-pattern are endless. When developing an application, designers and developers should try to empathize with mobile users, imagining themselves in various disparate scenarios: in the passenger seat of a car; carrying grocery bags; jogging on a treadmill; lying in bed; flying on an airplane; using the New York City subway system; having physical disabilities. Desktop and mobile applications differ primarily in their use cases. Accessibility is an important consideration in desktop software, but developers can rely on controlled environments and supportive technologies for the physically impaired. Mobile application developers should provide value for users operating in the least friendly conditions. The scope of progressive enhancement concerns must expand to include not only physical disability, but also environmental disability.

Sound Off

The direct ancestor of both the iPhone and the iPod Touch is the iPod. Apple has taken great care to ensure that both devices continue the legacy of the iPod by providing users with an elegant and familiar interface for playing audio. The interaction between applications, the operating system, and the iPod application has been carefully designed to allow interoperation without diminishing the value of the user experience.

The iPod application recognizes incoming telephone calls and fades to silence to let users answer and take part in conversations. When a call is disconnected, the iPod application fades the audio back in and resumes playback at the last known point in the track. If the iPhone receives a text message while the user is listening to music, the SMS alert sound plays along with the current song. All standard notification sounds work in the same fashion.

The sound off anti-pattern occurs when applications take exclusive control of the audio features of the device despite the preferences of the user.

Games are particularly guilty of implementing the sound off anti-pattern. Many games include rich sound effects and audio tracks. In an attempt to avoid auditory chaos, developers often give their applications exclusive use of the audio output mechanism, effectively silencing the iPod against the wishes of the user.

The Cocoa Touch audio frameworks allow developers to take control of the audio, but the practice is generally undesirable. Instead, developers should ask users for their preference. An excellent example of the user-centric approach is the game Flight Control. When the application launches, Flight Control asks if you wish to listen to your own music or listen to the audio provided by the game. (A third possible option would be to intermix the sounds, but for games with lots of sound effects, the result of mixing those sounds with the current song playing in the iPod application could be unpleasant.)

In this way, Flight Control offers two distinct, viable options. Most importantly, the game developers have shifted the power to choose from the application to the user. Figure 9-18 shows the sound management screen of the Flight Control game.

Figure 9-18. The sound management screen of Flight Control

Developers should consider the importance of the iPod application in typical mobile use cases and attempt to incorporate the user's desire to listen to music while playing games or using applications. Implementing exclusive locks on the audio output mechanism effectively removes the ability to use the iPhone or iPod Touch as a music player—a distasteful choice for many users.

Index

A

accelerometer, 137–138
accelerometer sensor, 10
accelerometer:didAccelerate: method, 138
accessoryView view (table cells), 121, 122
accuracy (touch), 62–68
action sheets, 152
active indicators (tables), 122
addTarget:action:forControlEvents: method,
    69, 90
alerts, 152
alphabetical lists, index ribbon for, 22
ambient light sensor, 10, 14
AmbientSound category, 144
anti-patterns, 147–165
    app as OS, 155–157
    billboards, 147–150
    bouncers, 157–159
    bullhorns, 152–154
    gesture hijacking, 160
    high bar, 163
    memory lapse, 161–163
    sleight of hand, 150–151
    sound off, 164
    spin zone, 157
app as OS anti-pattern, 155–157
App Store
    application categories, 19
    lazy loading, 130
    network failure notification, 154
    rejection from, 13
AppKit framework, 1
Apple HIG (see HIG; mobile HIG)
application data, saving, 126

application icon, size of, 15
application interaction patterns, 83–87
    combination interfaces, 87
    command interfaces, 83
    modal interfaces, 85
    navigation interfaces, 85
    problematic (see anti-patterns)
    radio interfaces, 84
application templates, 24, 27–35
    Core Data templates, 35
    for immersive applications, 25
    for light utilities, 25
    for productivity applications, 24
    view controllers, 29–34
application types, 19–25
    immersive applications, 25
    light utilities, 24, 94
    productivity tools, 20–24
applicationDidBecomeActive: method, 48
applications
    first experiences with, 17, 161
    launch images, 15, 44
    sharing data between, 54–55
    starting and quitting, feel of, 15
applicationWillResignActive: method, 47
applicationWillTerminate: method, 48
arbitrary shapes, interaction with, 74
archiving, classes for, 9
assisted scrolling, 22
audio functionality, hijacking, 164
Audio Queue Services, 143
audio support, 140–145
audio units, 143
AudioServicesPlayAlertSound function, 142
AudioServicesPlaySystemSound function, 141

We'd like to hear your suggestions for improving our indexes. Send email to *index@oreilly.com*.

authentication credentials, obtaining, 158
AVAudioPlayer class, 143

B

background processes, disallowed, 37
baseline features (see progressive
        enhancement)
battery level indicator, 154
billboard anti-pattern, 147–150
Birdfeed application, 85, 86
Black and White application, 153
bouncer anti-pattern, 157–159
bullhorn anti-pattern, 152–154
button controls, 91–98
    modal buttons, 98–100
buttons
    info ("i") buttons, 94
    physical (on device), 10
    size of (see fingertip size, designing for)
buttonWithType: method (UIButton), 91

C

caching before termination, 51
caching data, 126
    memory lapse anti-pattern, 161–163
    user input, 127
Calculator application (built-in), 24
canOpenURL: method, 54
categories of audio sessions, 144
CFBundleURLTypes key, 53
CGPoint values, 62
classes, 2
    control classes, 88–91
    control classes, standard, 91–114
    controller classes, 29
    Foundation class tree, 5
    registering with UIControl descendants, 89
    UIKit framework, 2–3
CLLocationAccuracy key, 136
CLLocationManager class, 134
CLLocationManagerDelegate protocol, 135
Clock application, 84
cloud-based processing, 55
Cocoa Touch, about, 1–10
    Foundation framework, 2, 4–9
    garbage collection, 9
    human interface guidelines, 13
    UIKit framework, 1, 2–3

Cocoa Touch applications (see entries at
        application)
collection management classes, 5
combination interfaces, 87
command interfaces, 83
connectivity (network), 126–133
    alerting users about, 128
    caching user input, 127, 161–163
    lazy data loading, 129–132
    maintaining state and persisting data, 126
    memory lapse anti-pattern, 161–163
    peer connectivity and GameKit, 132
connectivity (peer), 132
Contacts application
    modal buttons in, 98
    Ringtones screen, 106
Contacts application (built-in), 20, 33
contentView view (table cells), 121
control accessories (tables), 122
control events, 89
control types, 91–114
    buttons, 91–98
    modal controls, 98–100
    search bars, 109
    segmented controls, 111–114
    sliders, 103–104
    tables and pickers, 106–109
controller classes, 29
controllers (see view controllers)
controls, 88–91
    scrolling controls, 114–120
    segmented, 111–114
    tables and embedded controls, 120–123
    target-action mechanism, 89
    touchable, size of, 62–64
cooperative single-tasking, 17, 37–56
    custom protocol handlers, 17, 52–54
    handling interruptions, 47–48
    handling terminations, 47, 51–52
    launching quickly, 43–47
    push notifications, 55–56
    shared data, 54–55
Core Data
    lazy loading with, 131
    for saving application data, 126
    templates, 35
Core Location framework, 133–137
credentials, obtaining, 158
custom protocol handlers, 17, 52–54

D

data caching, 126
  memory lapse anti-pattern, 161–163
  user input, 127
data loading, lazy, 129
data persistence, when no connectivity, 126
data sharing between applications, 54–55
Default-scheme.png file, 44
Default.png file, 44
desiredAccuracy property, 136
detail disclosure buttons, 122
detail views (productivity applications), 23–24
detecting touch
  with arbitrary shapes, 74
  gesture hijacking, 160
  multiple touches, 70
  swipes and drags, 72
  taps, 68–69
  touch and hold, 70
device specifications, 10
didRotateFromInterfaceOrientation:fromInter
    faceOrientation: method, 140
disclosure buttons, 122
disclosure indicators, 121
dismissive gesture (swiping), 73
disruptive notifications, 153
distanceFilter property, 137
double taps, detecting, 69
Draggable class, 72
drags, detecting, 72

E

EAGLContext class, 34
EAGLView class, 34
embedded controls, 120–123
  active indicators and control accessories,
    122
  disclosure indicators, 121
enlarged virtual hit areas, 64
events (in response to touches), 60, 89
exit processes, managing, 51–52

F

Face Melter application, 84
Facebook application, 151, 158, 162
failure states, notification of, 16
file system classes, 7
filesystem IO, 16

fingertip size, designing for, 15, 23, 62–68
first impressions of applications, 17, 161
Flickr-like applications, 55
Flight Control application, 164
focus interruptions, handling, 47–48
Foundation framework, 2, 4–9
  garbage collection and, 9
Foursquare application, 84, 159
frameworks for Mac OS X programming, 1–9

G

GameKit framework, 132
games, 34
  audio hijacking, 164
  as immersive applications, 25
garbage collection, 9
gesture hijacking, 160

H

haptic feedback, lack of, 150
hardware feature abstractions, 4
hardware power button, 10
HIG (Human Interface Guidelines), 11, 13
  (see also Apple HIG; mobile HIG)
  deviating from, risk of, 13
high bar anti-pattern, 163
hitTest:withEvent: method, 64, 66
hold (after touch), detecting, 70
home button (depressible), 10
hot area for touches, 63
Human Interface Guidelines, 13 (see HIG;
    mobile HIG)

I

"i" buttons, 94
icon size, application, 15
image at launch (see launch image)
ImageSearch application (example), 38
  handling interruptions, 48
  handling terminations, 51
  launching quickly, 45–47
imageView view (table cells), 121
immersive applications, 25
  games, 148
  sound off anti-pattern, 164
index ribbon for logically ordered lists, 22
info buttons, 94
Info.plist file, 53

input caching
    memory lapse anti-pattern, 161–163
instant message client (example), 55
interaction patterns, 83–87
    combination interfaces, 87
    command interfaces, 83
    modal interfaces, 85
    navigation interfaces, 85
    radio interfaces, 84
interface controls (see controls)
interface interaction patterns
    problematic (see anti-patterns)
interfaceOrientation property, 139
interprocess communication (IPC) classes, 7
interruptions, handling, 47–48
IPC classes, 7
iPhone device, about, 10
iPhone Human Interface Guidelines (see mobile
        HIG)
iPhone SDK, 13
iPod application, 141
    sound off anti-pattern with, 164
iPod Touch device, about, 10
irregular shapes, interaction with, 74

K

kCLLocationAccuracy- values, 136
keyboard (onscreen), 64

L

landscape orientation, switching to portrait,
        139–140, 157
launch image, 15, 44
launch states, 15
launching of applications, 43–47
    example application, 45–47
    feel of, 15
    first-time experience, 17, 161
lazy data loading, 129–132
light sensor (see ambient light sensor)
light utilities, 24, 94
LinkedIn application, 158
LiveAudio category, 144
loading data lazily, 129–132
location awareness, 133–137
locationInView: method (UITouch), 59
locationManager:didUpdateToLocation:from
        Location method, 136

locking, classes for, 7
logical controllers, 4
logically ordered lists, index ribbon for, 22
login requests, managing, 158

M

Mac frameworks, 1–9
Mail application, 128, 155
mailto:// URL scheme, 155
maintaining state when no connectivity, 126
managed objects, 131
Maps application, 136
MediaPlayback category, 144
memory lapse anti-pattern, 161–163
mobile HIG, 12–18
    deviating from, risk of, 13
    progressive enhancement, 16
    seamless interaction, 15–16
    single user experience, 13–14
modal alerts, 152
modal controls, 98–100
modal interfaces, 85
    combined with other interfaces, 87
ModalButton class, 98–100
ModalButtonViewController class, 99
Multi-Touch interface, 57
multiple taps, detecting, 69
multiple touches, detecting, 70
multitasking, 37
    (see also cooperative single-tasking)

N

navigation controllers, 32–34
    sleight of hand anti-pattern, 151
navigation interfaces, 85
    combined with other interfaces, 87
navigation-based application template, 24, 30
navigation-based applications, 32, 33, 151
network communications, handling, 16
network connectivity, 126–133
    alerting users about, 128
    caching user input, 127, 161–163
    lazy data loading, 129–132
    maintaining state and persisting data, 126
    memory lapse anti-pattern, 161–163
    peer connectivity with GameKit, 132
    when poor, 17
New Project dialog (Xcode), 28

Notes application, 84
notification bullhorns, 152–154
notifications, classes for, 9
notifications on network connectivity, 128
NSArray objects, 2
    for saving application data, 126
NSDictionary objects, for saving application
        data, 126
NSObject class, 2
numeric badges, 55

O

Objective-C language services, classes for, 9
objects (touchable), 62–68
    overlapping views, 68
    shape of, 66, 74
    size of, 62–64
    view placement, 67
OmniFocus application, 85
OpenAL framework, 143
OpenGL ES application template, 25, 34
openURL: method, 52
operating system services, classes for, 7
orientation, rotating, 139–140, 157
overlapping views, 68

P

paging, 114
pagingEnabled attribute (UIScrollView), 114
partial rotation support, 157
passive indicators, 121, 153
password requests, managing, 158
peer connectivity with GameKit, 132
persisting data when no connectivity, 126
phase property (UITouch), 58
Phone application (built-in), 52
Photos application, 73, 114
picker controls, 106–109
placement of views, 67
    overlapping views, 68
PlayAndRecord category, 144
PNG file, as launch image, 15, 44
pointInside:withEvent: method, 64, 66
portrait orientation, switching to landscape,
        139–140, 157
power button, device, 10
precision of use, expectations for, 15

previousLocationInView: method (UITouch),
        59
productivity-focused applications, 20–24
    detail views in, 23–24
    scrolling in, 20–22
progressive enhancement, 11, 16, 125–145
    accelerometer support, 137–138
    audio support, 140–145
    high bar anti-pattern, 163
    location awareness, 133–137
    network connectivity, 126–133
    rotation support, 139–140
projects, starting new, 28
protocol handlers, custom, 17, 52–54
proximity sensor, 10
push notifications, 55–56

Q

quick-launching, 43–47
    example application, 45–47
quitting an application
    feel of, 15
    handling terminations, 47, 51–52

R

radio interfaces, 84
    combined with other interfaces, 87
RecordAudio category, 144
registering classes with UIControl descendants,
        89
registration management, 158
resolution, device, 10
responder chain, 59–61
REST over HTTP, 127
Ring/Silent switch, 141
rocker switches, 10
rotation support, 139–140
    partial, 157

S

Safari application, 155
sandboxes, 54
saving application data, 126
screen resolution, 10
scrolling
    controls for, 114–120
    in productivity applications, 20–22
seamless interaction, 15–16

search bars, 109

searchBar:textDidChange: method, 109

searchBarSearchButtonClicked: method, 109

searchBarTextDidEndEditing: method, 109

segmented controls, 111–114

sensors, device, 10

setMultipleTouchEnabled: method, 60

Settings application, 134

shape of touchable objects, 66

    arbitrary shapes, 74

sharedAccelerometer method, 138

sharing data between applications, 54–55

shouldAutorotateToInterfaceOrientation:inter
        faceOrientation: method, 139

Shovel application, 126, 155, 162

showsHorizontalScrollIndicator property, 114

showsVerticalScrollIndicator property, 114

signal strength indicators, 128

single taps, detecting, 68

single user experience, creating, 13–14

single-tasking (see cooperative single-tasking)

size, fingertip (see fingertip size, designing for)

size of application icons, 15

size of touchable objects, 62–64

sleep() function, abuse of, 148

sleight of hand anti-pattern, 150–151

slider controls, 103–104

snapshot, taking before termination, 51

SoloAmbientSound category, 144

sound (audio support), 140–145

sound off anti-pattern, 164

spin zone anti-pattern, 157

splash screens, 15, 44, 148

SQLite

    lazy loading with, 129

    saving application data with, 126

starting applications

    feel of, 15

    first-time experience, 17, 161

    quickly, 43–47

state maintenance, 15, 126

status bar, when open, 48

status notifications, 16

    push notifications, 55–56

Stocks application, 154

Streaks application, 148

stretchable images, 92

stretchableImageWithLeftCapWidth:topCap
        Height: method, 92

string classes, 5

subordinate views, 24

success states, notification of, 16

swipe-to-delete gesture, 73

swipes, detecting, 72

    gesture hijacking, 160

System Sound Services, 141

T

tab bar application template, 24, 25, 31

tab-based applications, 31

table controls, 106–109

tables, 120–123

    active indicators and control accessories,
        122

    disclosure indicator, 121

tableView:accessoryButtonTappedForRowWi
        thIndexPath: method, 123

tactile feedback, lack of, 150

tapCount property (UITouch), 58, 69

taps, detecting, 68–69

    gesture hijacking, 160

target-action mechanism, 89

task management (see cooperative single-
        tasking)

TED application, 87, 161

tel:// protocol handler, 52

templates for applications, 24, 27–35

    Core Data templates, 35

    for immersive applications, 25

    for light utilities, 25

    for productivity applications, 24

    view controllers, 29–34

terminations, handling, 47, 51–52

themes (design), 149

Things application, 84

threading classes, 7

3D games, 25, 34

thumb, slider, 103

TileSearchViewController class, 109

timer, for handling touch and hold, 70

timestamp property (CLLocation), 137

timestamp property (UITouch), 58

touch and hold, detecting, 70

touch patterns, 57–74

    accuracy considerations, 62–68

    arbitrary shapes, 74

    detecting multiple touches, 70

    detecting taps, 68–69

detecting touch and hold, 70
gesture hijacking, 160
responder chain, 58–61
swipes and drags, 72
touchable objects, 62–68
overlapping views, 68
shape of, 66, 74
view placement, 67
touches argument (touch event handlers), 70
touchesBegan:withEvent: method, 60, 69
touchesCancelled:withEvent: method, 61
touchesEnded:withEvent: method, 61, 99
touchesMoved:withEvent: method, 60
touchesMovied:withEvent: method, 72
track, slider, 103
tree-like graphs, 20
Tweetie application, 31, 147, 153, 160, 162
Twitterrific application, 149

U

UIAccelerometer class, 138
UIAccelerometerDelegate interface, 138
UIAlertView class, 152
UIApplication class, 54
UIApplicationDelegate protocol, 47
UIButton class, 91–98
info ("i") buttons, 94
UIButtonType key, 91
UIButtonTypeInfoDark class, 96
UIButtonTypeInfoLight class, 96
UIControl classes, 88–91
standard control types, 91–114
target-action mechanism, 89
UIControlEvent class, 89, 103
UIControlEventTouchUpInside event, 99
UIControlEventValueChanged type, 103
UIControlStateHigh state, 92
UIControlStateNormal state, 92
UIDatePickerModeCountdownTimer class, 107
UIDatePickerModeDate class, 107
UIDatePickerModeDateAndTime class, 107
UIDatePickerModeTime class, 107
UIEvent class, 60
UIImageView objects, touch-enabled, 67
UIKit framework, 1, 2–3
UINavigationController class, 85
with navigation-based applications, 32
UIPickerView class, 106–109

UIResponder class, 2–3, 59
UIScrollView class, 114–120
UISearchBarDelegate class, 109
UISlider class, 103–104
UITabBar class, 84
UITabBarController class, 30
with tab-based applications, 31
UITableView class, 20, 106–107
index ribbon for alphabetical lists, 22
UITableViewCell class, 121
UITableViewCellAccessoryCheckmark key, 122
UITableViewCellAccessoryDetailDisclosureBu
tton class, 122
UITableViewCellAccessoryDisclosureIndicato
r class, 122
UITableViewController class, 30, 33
UITableViewControllerCell class, 33
UITableViewDelegate class, 120–123
UIToolBar class, 83
UITouch class, 58–61
UIView class, UIViewController and, 29
UIViewController class, 29
in modal interfaces, 85
in navigation interfaces, 85
rotation support, 139
with view-based applications, 30
URL handling classes, 7
URLs, custom, 17, 52–54
user experience (UX), 10, 14
anti-patterns, 147–165
guidelines (see HIG; mobile HIG)
interaction patterns (see application
interaction patterns)
touch patterns (see touch patterns)
user input, caching, 127
memory lapse anti-pattern, 161–163
user interface, importance of, 2
user interface controls (see controls)
user interface interaction patterns, 83
combination interfaces, 87
command interfaces, 83
modal interfaces, 85
navigation interfaces, 85
problematic (see anti-patterns)
radio interfaces, 84
UserInterfaceSoundEffects category, 144
username requests, managing, 158
utilities (light), 24, 94

utility application template, 25, 31, 94
UX (see user experience)

V

value objects, 4, 5
vertical scrolling (see scrolling)
vibration, triggering, 141
view controllers, 29–34
view placement, 67
view property (UITouch), 58
view-based application template, 25, 30
view-based controller template
    example application (ImageSearch), 38
views, overlapping, 68
virtual controls, 150
virtual hit areas, 64
    placement of, 67
volume rocker switch, 10, 141

W

Weather application, 24, 95
willAnimateFirstHalfOfRotationToInterfaceO
        rientation:duration: method, 140
willAnimateSecondHalfOfRotationToInterfac
        eOrientation:duration:, 140
willRotateToInterfaceOrientation:toInterface
        Orientation:duration: method, 140
window property (UITouch), 58
window-based Core Data application template,
    35
WordPress application, 158
Wurdle application, 148

X

Xcode templates, 24, 27–35
    Core Data templates, 35
    for immersive applications, 25
    for light utilities, 25
    for productivity applications, 24
    view controllers, 29–34
XML classes, 5

Z

z-axis, views overlapping on, 68

About the Author

Toby Boudreaux has been developing for Mac OS X using Objective-C and Cocoa since 2000. He has spoken at WWDC on the topic of Hybrid Cocoa/Web applications—a very relevant topic for the iPhone. He is the CTO of The Barbarian Group, an interactive/software shop based in the U.S. He focuses evenly on OS X/ iPhone application development and Web development, and acts as a mentor to his team, liaison to his clients, and representative to the community. Toby has authored and acted as technical editor for books and articles related to programming, and he specializes in web development for consumer markets using open technologies and in Mac/iPhone development.

Colophon

The animal on the cover of *Programming the iPhone User Experience* is a six-shafted bird of paradise (*Parotia sefilata*), also known as a Western Parotia or six-wired bird of paradise. The male of this species is 13 inches long with a short bill and black iridescent plumage that varies between bronze, green, and purple. It has golden feathers on its breast and silver feathers on its crown. From behind each eye spring the head wires— each six inches long with a small oval tip—that give the bird its name. On the sides of its breast are black plumes that, when elongated, make the bird appear to be double its real size. As with most birds of paradise, the female is brown and unadorned.

Native to Indonesia, the six-shafted bird of paradise inhabits the mountain forests of Western New Guinea, living on a diet of fruits and figs. During courtship, the male performs an elaborate dance by shaking its head with its black plumes elongated to show off its adornment. This bird is common throughout its range and is not considered a threatened species.

The cover image is from Cassell's *Natural History*. The cover font is Adobe ITC Garamond. The text font is Linotype Birka; the heading font is Adobe Myriad Condensed; and the code font is LucasFont's TheSansMonoCondensed.

Related Titles from O'Reilly

Macintosh

AppleScript: The Definitive
Guide, *2nd Edition*

AppleScript: The Missing Manual

Appleworks 6: The Missing
Manual

The Best of the Joy of Tech

FileMaker Pro 8: The Missing
Manual

FileMaker Pro 9: The Missing
Manual

GarageBand 2:
The Missing Manual

iBook Fan Book

iLife '05: The Missing Manual

iMovie 6 & iDVD:
The Missing Manual

iPhoto 6: The Missing Manual

iPhoto '08: The Missing Manual

iPod: The Missing Manual,
6th Edition

iWork '05: The Missing Manual

Mac Annoyances

Mac OS X Tiger Pocket Guide

Mac OS X Leopard Pocket Guide

Mac OS X: The Missing Manual,
Tiger Edition

Mac OS X: The Missing Manual,
Leopard Edition

Mac OS X Power Hound,
2nd Edition

Mac OS X Unwired

Modding Mac OS X

Office 2004 for the Macintosh:
The Missing Manual

Office 2008 for the Macintosh:
The Missing Manual

Revolution in The Valley

Switching to the Mac:
The Missing Manual,
Leopard Edition

Mac Developers

Building Cocoa Applications:
A Step-By-Step Guide

Cocoa in a Nutshell

Essential Mac OS X Panther
Server Administration

Learning Carbon

Learning Cocoa with
Objective-C, *2nd Edition*

Learning Unix for
Mac OS X Tiger

Mac OS X for Java Geeks

Mac OS X Panther Hacks

Mac OS X Tiger in a Nutshell

Mac OS X Tiger for Unix Geeks,
4th Edition

Objective-C Pocket Reference

Running Mac OS X Tiger

Try the online edition free for 45 days

Developing and Designing Cocoa Touch Applications

Programming the
iPhone User Experience

O'REILLY®

Toby Boudreaux

Get the information you need when you need it, with Safari Books Online. Safari Books Online contains the complete version of the print book in your hands plus thousands of titles from the best technical publishers, with sample code ready to cut and paste into your applications.

Safari is designed for people in a hurry to get the answers they need so they can get the job done. You can find what you need in the morning, and put it to work in the afternoon. As simple as cut, paste, and program.

To try out Safari and the online edition of the above title FREE for 45 days, go to www.oreilly.com/go/safarienabled and enter the coupon code TMDYYBI.

To see the complete Safari Library visit:
safari.oreilly.com